SHOW ME A FORESTER

By

Josh Shroyer

Words Matter Publishing
P.O. Box 1190
Decatur, Il 62525
www.wordsmatterpublishing.com

ISBN 13: 978-1-968542-20-7

Library of Congress Catalog Card Number: 2026933623

Dedication

This book is dedicated to my wife, without you, none of this journey would have been possible.

Acknowledgments

I would like to acknowledge all the friends, family, fellow foresters, and mentors, not only from my career, but from my life, who have been a part of this wonderful journey. A special thanks to a good family friend, Paige Schooner, who provided her artistic talents in the form of sketches for places I had no photos; you really filled in the gaps nicely. To Tammy and the whole Words Matter Publishing crew, who have been so great and easy to work with, especially on this next book. To Larry Lackamp, my first "boss" who invested so much in me and was such a valuable teacher so early in my career. To my family for all of the continued support. To Nate, who loves to work beside me and help continue the search for more adventures, and most importantly, to my wife, Tammy, who has continued being my "ride or die" partner on this crazy journey we call life.

Table of Contents

Dedication iii

Acknowledgments v

Prologue ix

Chapter 1: Trees of My Youth............................... 1

Chapter 2: College Antics.................................... 9

Chapter 3: Slashing Projects 13

Chapter 4: Summer Camp 17

Chapter 5: My First Conservation Job................. 21

Chapter 6: Peak Fall Color 27

Chapter 7: Don't Fall Asleep.............................. 33

Chapter 8: Busted Ankle 37

Chapter 9: Flooded Inventory on the Platte 43

Chapter 10: Blown Tires on the Jeep 47

Chapter 11: Hole In the Wall Eateries and Peanut Butter Pie ... 53

Chapter 12: Eerie Days in the Forest 59

Chapter 13: Forestry and FFA 67

Chapter 14: From Planting to EAB, Nice Ash 73

Chapter 15: The Long and Short of Flagging 79

Chapter 16: Don't Touch My Snacks 85

Chapter 17: The Redwood Healing............................... 89

Chapter 18: Hey Bear, (and Wolves and Cats, oh my!) 95

Chapter 19: Technology 101

Chapter 20: Basketball Anyone? 107

Chapter 21: One More Plot 113

Chapter 22: My Own Schedule................................. 119

Chapter 23: Scout Camp...................................... 127

Chapter 24: No Typical Day 135

Chapter 25: I Found Nemo.................................... 141

Chapter 26: Special Moments 145

Chapter 27: Champion Trees.................................. 155

Chapter 28: Walnut 161

Chapter 29: Dream Interpretation.............................. 171

Chapter 30: Tree Marking Gloves.............................. 173

Chapter 31: How Far Back is Backcountry?..................... 177

Chapter 32: A Forester Wife, A Forester Life.................... 185

Chapter 33: Lunch Spots...................................... 191

Chapter 34: Fascinating Finds in the Forest..................... 195

Chapter 35: The Englewood Project............................ 199

Chapter 36: Here Comes Chaos............................... 203

Prologue

I think you need to read this! This important information will help bring some context to this book. *Show Me a Forester* is the second book in a series that doesn't yet have a final number of books determined. However, it is a logical second step for me. The first, *Show Me a Firefighter*, focused on stories from the wild-fire part of my career, and now this will focus more on the forester part of my career.

The format will be very similar to the first book, with short stories and a bit of humor injected. Humor seems to be the thing that I have enjoyed the most about writing. No, humor can't be in everything I write, but I will always try to come back to that. It boils down to a very basic fundamental belief of mine that we need to spread joy and happiness in this life, and if I can tell a story that is funny and bring a smile or even a laugh, then I feel I have been successful.

With that, I feel I need to tell of a time when humor was absent. Without a huge backstory, I will just say that we lost our youngest child, Drew, at age 17, a number of years ago. Drew was a pun master and had the best sense of humor of any kid I had ever known. He was always cracking a joke, making people smile and laugh, and really enjoyed getting into pun wars with anyone who would take the time to try their luck against his well-developed skill.

Well, as you can imagine, that loss took its toll. Humor was the last thing I could think about, and the ability to write my stories, or write at all for that matter, completely vanished. Fast forward a number of years, and I was in a bookstore just casually browsing some books about trees, yeah, imagine that. This was more to pass the time while I waited for my wife, Tammy, and son, Nate, while they actually shopped. While I was looking at one particular book, I heard a familiar voice. Not out loud, not even in my head, but deep down in my heart. He said, "I think it is time to get back to writing your stories, Dad. It has been long enough."

Now, I don't recall there being that much dust in that bookstore, but some must have gotten in my eyes. However, that moment changed something in me. I think I even casually mentioned to Nate that I was going to start on my stories again. With that, my writing came back to life, and I was able to find my lost humor. Yes, many of my stories were written years ago, but others were just a collection of notes, and I still needed to develop the story. With renewed energy, I started not only developing those stories from notes, but even spent time remembering other fun times, and started getting those stories down on paper. Or, well, on the computer anyway.

Now, at this time, I want to also mention another person whose humor I have always enjoyed. I met Terry Gordan while in forestry school. He had a well-developed sense of not only humor but comedic timing as well. While we worked for the Missouri Department of Conservation at the same time, we were in different parts of the state and only occasionally met up, usually at a training or meeting. Regardless, I always looked forward to his jokes, humor, and wit.

Time and distance, and differing jobs, led us apart, but we eventually reconnected. Terry reminded me that there are no coincidences, only divine intervention. One of my regrets in life is that Terry and Drew never met each other. I think they would have genuinely enjoyed each other's company as they had very similar humor styles. The more I think about it, the more I can see similarities. I think Drew may have met his match in Terry, or maybe it would be Terry meeting his match in Drew.

When I finally told Terry what had happened in that bookstore and that I was writing again, he again reminded me of no coincidences. He celebrated and encouraged me in a way that only he could have. These days, I always look forward to meeting up with Terry and his family, as you need to surround yourself with good people all the time. Terry is one who likes to spread joy and happiness. He's an avid reader and has a habit of giving, or even buying and sending books to others who need encouragement, or just a little extra something. I hope that my books can be one of those that get sent to someone who needs a smile or a laugh at just the right moment.

If that happens for even one person, I truly have Terry and Drew to thank for that. Each in their own respective way. So, hopefully, as you are reading book number two, you are looking forward to learning about this small-town boy from Missouri as I tell stories of becoming a forester. There are even some more recent stories about our family consulting business. I continue to show who I am and things I am still learning in forestry and in life. One thing I feel confident of, I will continue to write stories. I will continue to look for the humor in it all, and I will continue to share those stories as a way to spread joy and happiness.

Whether all those stories get published or not, only time will tell. The writing has been both fun and therapeutic, so I will keep riding the wave. So, with that, please sit back and enjoy this next collection. Hopefully, you will laugh a bit and also think back to some of your own tree stories and memories. I think we all have more in common than we realize, no matter where we are from or where we walk through the trees.

Chapter 1

Trees of My Youth

Trees made an impact on me before I decided I wanted to be a forester. Maybe I knew much earlier than I originally realized, or maybe it was just the right influences that led me down a wooded path. Regardless, some of my vivid early memories center on a number of trees.

When I was born, we lived in the small rental house my grandparents owned in Wellington, but soon after, Mom and Dad moved to a house in Napoleon. We only lived there until I was 5, but a pair of trees filled my mind with adventure. There were two sweetgum trees in the front yard, and with help from Mom or Dad, I was able to get to the lowest limbs. I sat on the limbs or hung from my arms, swinging around. As I got a bit taller, I was able to jump and reach them on my own, and that allowed me to start climbing higher when neither was watching me. I would throw the seeds, otherwise known as the "spikey balls," at birds or the tree itself, imagining I was a pitcher for the Kansas City Royals.

Our neighbors to the West of our house had kids, and I started to play with them. Although I don't know what kind of tree it was, there was a tree with a swing tied to a branch that we played on a lot. I thought it was cool to have a swing in a tree. The only other swings I knew were built by my Grandpa Jack. Anything my grandpa built was built tough. It was welded, three-inch steel pipe. It lasted me and my two younger brothers, and I even had it for my boys growing up. My grandparents would never have allowed a swing to be tied in a tree, so this was just fascinating for me.

As my mom was a teacher in Wellington, I rode to Wellington with her each day and got dropped off at a babysitter. The two things I remember learning from this place were my first exposure to coffee, well, a little coffee and lots of milk to dunk my cookies in, and learning what maple seeds were like. There was a large silver maple in their front yard, and the "helicopters," as I called them, were fun seeds to play with as I grabbed a handful and tossed them as high as I could. They would then twirl and spin on their way back down.

After we left Napoleon and moved back to Wellington, we had the house right across the street from my grandparents. This allowed me to get a lot more tree exposure. I'll start with our house first. In the back, at the edge of the yard, were lots of tall, slender trees that each spring blossomed and smelled wonderful. I would learn these as black locust trees. As delightful as they smelled, they also had a dark side. Mainly thorns on the smaller branches. Only in college would I learn that the technical term for them is stipular spines. At the time, I just knew to be careful in grabbing a twig or small branch.

I had already learned about silver maple, and we had one on the property line between us and the neighbor. As it could have been the neighbor's tree, I didn't climb it like I would have liked, but it was still fun to grab up a handful of seeds and toss them.

Across the street at my grandparents, the real tree learning began. They had quite an assortment of trees, and looking back, I didn't realize how fortunate I was at the time. Through the decades, my grandparents had cultivated a diverse yard. First, there were the walnut and pecan trees that Grandpa Jack cared for so he could collect the nuts to crack and eat. Even as a kid, I was taught the difference between the two without relying on the nuts. The differences in the bark texture and subtle leaf characteristics were fascinating.

The fruit trees were my favorite, of course. They had a couple of apple trees and a cherry tree. My grandma would use these to make apple or cherry pies, cherry jelly, or other desserts. However, my absolute favorite was the apricot tree. Each summer, I would get in trouble because I would try to knock the apricots down so I could eat them before the ants got to them. Otherwise, I had to sit and wait for them to fall. I would then pounce on them and eat them as fast as I could. Funny, I don't remember ever getting any apricot jelly or apricot preserves. I can't imagine that I ate so many that there wasn't enough left. Oh no, not me.

My grandparents also had some holly trees near the lower drive. These stayed green all year long and had prickly leaves and red berries. Only in college would I learn that they were not native to that part of Missouri, but growing up, I just thought they were normal, and everyone may have them. With the holiday-themed red and green of the tree, my mom and grandma would make holly wreaths from selected trimmings of the branches every Christmas.

There was a tree next to their house that had thick, tough, almost leathery leaves that I would learn was a magnolia tree. I was fascinated by these leaves as they were bigger than any other tree around, and they were just so thick and tough-feeling. I could only pick up the ones that had dropped. I didn't know it then, but again, a tree that was not common, and they were afraid of me pulling leaves from that tree.

One of my favorite climbing trees was right out the back door of their house and next to the patio. This was a large redbud tree. At the time, I just knew it as a good climbing tree, but now, I wonder if it could have been a state champion at some point. It was multi-stemmed, of course. As I got a bit bigger and older, I could climb pretty far up and even got as high as the roof peak of the 2-story house. I was always climbing that tree and only had a "few" accidents. One in particular scared me pretty good. I fell out of the tree backward somehow and landed flat on my back. I don't recall my head hitting, but the impact knocked every ounce of breath out of me!

Try as I might, I couldn't get air back in my chest, and I started to panic as any eight-year-old might in that situation. Eventually, my body relaxed, and I was able to breathe again, but that was not fun.

During another fall from that tree, I did hit my head. This time was a face-first plummet, and I only got one arm out to brace myself. I hit the concrete and scraped up my hand and elbow pretty badly, and I still don't know how I managed to avoid busting my nose or breaking teeth, but I had a huge goose egg bump on my forehead right above my eyebrows. I was probably about eight or nine at the time, and I was more worried about getting in trouble than anything, so I managed to hide in other parts of the yard for the rest of the day. I didn't want to get grounded from climbing. You know, kid priorities. I mean, I probably would have been told to rub some dirt on it anyway and suffer the consequences of my actions, but just the thought of getting banned from the tree would have been a far worse fate.

I learned about the state tree, the flowering dogwood, as there was one in their yard, as well as their neighbor's yard. I was fascinated with the dogwood flower and the red berries that it produced. I will say that the bright red berries were hard to resist early on, but I would not recommend eating them. Trust me, a bit bitter up front and sour on the stomach later. I learned quickly to stick with apricots and raspberries.

Up the hill, the neighbor to my grandparents had a large eastern white pine in her backyard near the alleyway. This was a great tree for climbing, and I can only imagine exactly how high I went. I do know one day it got a bit windy, and I was near the top. The tree was swaying more than I was ready for, and I froze up a bit. Problem was, no one knew I was up there, and even if I tried to shout, the noise of the wind would have drowned out my cries anyway. It was just time to figure it out myself. Afraid the movement of the tree and branches would cause me to lose my footing or grip, I stayed as close to the main trunk as possible.

This is also where most of the oozing sticky sap was located. I carefully inched my way down, branch by branch, until I got to the bottom. By then, I was sticky all over my hands and arms, and even my shoe laces were covered and sticky. I avoided that tree for a few weeks and especially on windy days, for the rest of the summer.

There were a handful of other trees right there close, but looking back, the trees I mentioned are the ones with the most memories. Who knows, maybe I was destined to be a forester and work with trees from day one. Even now, I will see a tree and think, "Oh, that looks like a good climber" as I see some good branch structure or notice how a tree looks similar to one of those I grew up with, and the memory floods back in.

Of course, as I got older and was in high school and college, I kept learning about those trees of my youth. I just find it fascinating that so many early memories are centered on some of those trees and how much I was taught about them. I planted a few trees in the limited space we had at our Wellington house. Some I grew from nuts from the pecan and walnut trees across the street from my grandparents, another one was from a seedling, a nice red oak that took off beautifully. Sadly, someone came along and cut it off when it was about 12 feet tall and slightly smaller in diameter than a paper towel roll. I was upset and wondered why anyone would do that. Thankfully, it did sprout back and grew into a wonderful shade tree.

I did once visit the owners of the Wellington house I grew up in and was able to look at that oak tree and the remaining walnut and pecan in the back yard. They are all doing rather well. Not good climbers as they have been pruned of their lower branches, but they are providing a great deal of shade for the house.

Nowadays, I still like to see a swing in a tree, but I also look to give some advice on how to do that, AND protect the tree from unnecessary injury. I still wonder in amazement at the random trees, totally out of their area and should not survive, but seemingly doing just fine, and think fondly of that magnolia and those holly trees. I still like to grab a handful of maple seeds and toss them in the air and watch them race to the ground, twirling all the way. I don't throw the sweetgum balls as much as I used to, but I can't help but think back to those first memories of those two trees in the front yard every time I see a sweetgum.

Speaking of those sweetgums, I did make it back to the Napoleon house to visit those trees recently. They are doing very well and have really grown. Telling the current owners of my stories of those trees and of my climbing exploits really opened the floodgates of early memories. It was nice to compare the things that seemed to change so much from my youth against the things that hadn't changed a bit, like the house still has the same kitchen cabinets and wallpaper that I remember from all those years ago. The trees? Those have definitely changed. Almost as if we had grown up together.

Photo of me while taking a break from climbing one of the sweetgum trees in my front yard in June of 1979. Photo by my mom, Kathy Shoyer.

Another Forrest Gump moment for me. Caught climbing some random maple in 1980. Photo by my mom, Kathy Shroyer.

Visiting the same sweetgum tree I used to climb and seeing how big it had grown. Photo by Nate Shroyer.

Chapter 2

College Antics

While attending the University of Missouri at Columbia to pursue and earn my forestry degree, I took every opportunity to not only earn a little extra cash, but also gain experience as well. This involved fighting wildfires over spring break and on weekends, or working in the research lab, cracking, separating, and weighing walnuts for a research project that was looking at the best varieties of walnut for nut meats.

One time, the University Extension Forester, Jack Slusher, came by the Forestry Club room asking for a number of students to help with a timber sale marking project. This was to be on the Wurdack Farms property in Southern Crawford County. As I was open to any chance to gain some practical experience and earn some money, I was one of the first to sign up.

On the weekend in question, the group of students all gathered up and piled into some of the University vans to make the two-and-a-half-hour drive to the work location. We were given the marking prescription and quart cans of tree marking paint, and were divided up. Now, up to this point, I had never used a paint gun or marked any trees. It is not that difficult, and with a little practice, I was up and running pretty well.

I don't really remember the marking scheme or any difficulties from that weekend, other than we marked a lot of black oak. Even if I do remember who was with me, I won't mention any names so as to keep them innocent for what I am about to relate. As young people may be prone to do, shenanigans ensued. Normally, when you are marking trees to be cut or harvested, you will put a slash or even a band of tree paint on the trunk at about head high, along with a spot of paint on a protected part of the lower stump. This lower spot is to remain once the tree has been cut and the paint slash is gone with the log. You can walk around and look at stumps, find the stump spot, and know if a tree was supposed to be cut or not based on the presence or absence of the stump mark.

Being unsupervised and maybe having a short attention span, I may have migrated from slashes on the trees to smiley faces. Of course, kids being kids, some of the others may have also started adding creative designs on the trees. By the end of the weekend, there was all manner of artwork and designs on the trees across the harvest project. I couldn't tell you how much extra tree marking paint was wasted on our creativity. Art or not, it was too much. Tree marking paint is much more expensive than the regular paint you buy locally from the hardware store.

I wondered all the way back that evening if someone would spill the beans as to our antics and what the resulting punishment would be. While I never did hear anything about that particular project, I look back now and think that maybe that wasn't my best professional action. I wonder what the loggers were thinking when they not only bid on that project, but were actually cutting it.

I realized, soon after, that we have an obligation to uphold professional standards. Regardless of being a forestry student, I still represented not only the University, but the Extension Forester, and the overall forester profession as well. This was an uncomfortable internal lesson I learned early on that has stayed with me through the rest of my career. Even today, I think about this lesson as I try to be as professional as possible, not only for myself, but for those around me and the forestry profession as a whole. College antics, while sometimes fun or funny at the time, need to be tempered. Thankfully, this was before cell phones and social media, so there is no proof, just my integrity to tell this story on myself and the lesson I learned from it.

Thankfully, other college antics were not quite as embarrassing or unprofessional. Some can hardly be called antics, even. Take the dates we took to The Bur Oak. Yeah, you are probably familiar with the Champion Bur Oak near McBaine. As a forestry student and tree guy, and broke to boot, that is where Tammy and I ended up on a number of our dates after we grabbed a $0.39 cheeseburger from McDonald's. Yep, we really were party animals back then.

One particularly fun adventure was the Forestry Conclave, where we all traveled to Minnesota and participated in forestry and logging skills like crosscut, speed chop, or the tobacco spit. I was never one for any sort of tobacco, much less the mouthful of brown saliva needed for the spit part of the competition. It was rather fun to watch some of the contestants turn green or even puke as it became too much for them. I think one guy even found his future wife when she out spit him. I think it was love at first dribble. Hey, I don't judge.

For me, the speed chop was less speed and more gasping and wheezing, but I did get through that pine cant. Eventually, I was beginning to think those squared-off logs were called cants because I can't chop through it! Tammy and I did the Jack and Jill log roll, and overall, a great time was had. Of course, there always had to be some sort of drama at these types of events, and overnight, some school took off with the giant Paul Bunyan statue. The only thing I know for certain is that it was not our school. To this day, I will blame, um, on second thought, I'll just keep that to myself.

One of our dates to The Bur Oak. I probably shouldn't have been up on the big rock, but I was. Photo by Tammy Shroyer.

Looking after an oak seedling I planted in Wellington.
Photo by Kathy Shroyer

Chapter 3

Slashing Projects

On the theme of gathering experience and making some extra money, there were other opportunities for both. As I was a transfer student, I was still getting to know some of the other students and assessing what their skill sets were. Some of them had connections with the Mark Twain National Forest and had bid on some timber sale slashing projects. This is where a timber sale has been completed, and now all the damaged trees need to be cut, and some of the residual trees that still need to be removed.

I was asked if I wanted to be part of the crew and if I had any chainsaw experience. I had run a chainsaw for years and had run my own firewood business in high school, and I figured that more than qualified me, so I said yes. With that, the next weekend, I went home and gathered up my chainsaw and tools, and headed back to college. The following weekend, we all loaded up in the trucks of some of the older students and headed out.

Most of the cutting we had to do was with smaller-diameter trees that had been damaged or had broken tops. You don't want to leave those as they will never amount to a good tree, and they take up valuable sun that could benefit a newly sprouted tree. This cutting was pretty easy and straightforward. Another requirement was to cut up tree tops that were hung up in other trees. This started to become some pretty technical cutting. The trees were under tension, and the branches were not always clear. Slow cutting and reading the tree were critical skills, as tension caused trees to want to roll or twist. Finally, the last requirement was cutting and releasing the spring poles. A spring pole is a tree that was bent over completely by another tree as it was cut and felled. The stress and tension in spring poles are extreme, and advanced cutting techniques should be employed to release them safely.

Keep in mind, we were a bunch of college students who really had no formal training with a chainsaw and no practical experience. Only later in my career would I be taught those advanced techniques and gain valuable practical experience to be able to not only safely, but efficiently mitigate trees in weird stresses, binds, and twists. But, for this cutting project? We were about to find out the hard way and learn many painful lessons.

I will consider myself one of the lucky ones; I only had some minor scrapes and incidents. As I was cutting one particular spring pole, it snapped sideways a bit and narrowly missed my face. It did, however, take my hard hat off and fling it about 20 yards down the slope. Embarrassed, I looked around to see if anyone had witnessed this. Everyone within sight was actively cutting their own trees, so I thought I was in the clear. I mention the hard hat as it was about the only piece of safety gear we had that day, other than leather gloves and some cheap logger-style boots. No safety chaps, no earplugs, no safety glasses, although I was wearing my regular eyeglasses and figured that was protection enough.

Another close call I had was learning about kickback on the chainsaw. Again, this was way before any formal training I ever had. Regardless, I was cutting up a tree's top to reduce its overall height and couldn't see the tip of my bar. I must have hit another branch, and it kicked violently back and hit my boot. I stopped and took inventory of all my toes! Yep, I still had all ten little piggies. However, the toe of my boot had a rather large notch missing from the front of the hard sole. These were two close calls in only a matter of hours on the first day, so I figured I had better take a good, long water break and collect myself.

It was during this break that I saw some of the other guys learning some of their valuable lessons. I was learning by seeing their misfortune as well. Again, I will not mention any names here so as to protect their pride and dignity. I saw one guy cut a spring pole, and when it snapped, it caught him square in the chest and flung him about 10 feet. Based on his reaction and struggling after, I'm pretty sure it knocked the wind out of him pretty hard, as well as scared the crap out of him to boot. I saw another guy have the top of a broken tree come out and hit him on the head and knock him to the ground.

At breaks and lunch over those 2 days, I also observed jeans that were cut and wondered how they didn't also cut their legs. Lucky, I guess. I saw gloves ripped out and counted two saws that were literally destroyed. Busted up so bad, I don't even think they could be used for parts. Again, I don't really recall anyone telling of their misadventures. I think they were all just as embarrassed as I was about these sorts of mishaps.

At the end of the trip, we loaded up and headed back to campus, beaten, bloodied, bruised, and feeling extremely lucky to have survived it all. Also, I don't recall ever hearing about any students bidding on slashing projects ever again. If they did, maybe I just blocked it out and chose not to participate. None of us ever really talked about that weekend after we were done, so I can only guess they were just as shocked and confused as I was. We were definitely humbled by the experience. What can I say, we were typical Gen X kids, we didn't let lack of instruction stop us from our life goals, we figured it out along the way. I would keep these lessons and my ten piggies with me the rest of my career. When safety is concerned, it is not the time to be a slow learner.

Of course, the lessons I learned on that particular trip stayed with me. I learned I didn't know as much as I thought I knew about chainsaws and working in the woods. I learned that safety equipment has a place, and I was now properly motivated to wear it correctly. I learned that it is ok not to divulge all our embarrassing predicaments. And I learned it is ok to not call others out on their mishaps, lest you get called out too.

A tree that probably should have been slashed down a long time ago. However, it did turn into a rather interesting goal post. And, if you look at the right side, it kinda looks like a dragon. Photo by Nate Shroyer

Chapter 4

Summer Camp

During my time at MU, part of the forestry curriculum was to attend a six-week summer camp deep in the Ozarks at University Forest. These six weeks were met with mixed emotions as, first and foremost, I not only would not be able to make money for those six weeks, but it also precluded any type of full-time summer work, as there was just not enough summer left after camp. This was a double whammy: less income, more tuition bills. On the other hand, the whole purpose of Summer Camp was to expose students to actual field work and experience real-world scenarios instead of just theory in the classroom.

I was looking forward to what we would learn while there, so I decided to make the best of the situation. The accommodations at camp were pretty top-notch if you ask me. There were a number of cabins that slept six or eight and had a common living area up front. Due to the remoteness of the camp, there was a staffed kitchen that made all our meals, and we ate in a dining hall.

The camp even had its own mascot of sorts. I don't remember his name, but there was a dog who hung around and acted like he was one of us. Instead of the regular old dog collar, he usually wore a red bandana, although I vaguely remember one week it was blue. I guess he was just trying to mix it up. You know, keep up with the latest Ozark backwoods fashion sense, or maybe to keep us guessing. Regardless, I do wonder what he thought of a new group of best friends that showed up each summer for six weeks and then vanished. He probably had a lot of stories to tell.

Due to these wonderful accommodations, most students stayed over the weekends for the entire six-week period. I, on the other hand, had important things to do on the weekends. I would say it was planning my upcoming wedding, but the reality was I was supporting Tammy as she did most of the planning and heavy lifting. I would show up late Friday night and spend Saturday and most of Sunday helping with the final decisions. Then Sunday evening, back to camp.

I was fortunate to have a fellow student and good friend, Jennifer, who was also heading back every weekend, so it made ride sharing easier, someone to not only share gas money but conversation as well.

Although every student who went to forestry school was taught this as common knowledge, a quick check of the University's web page will tell us that University Forest was originally endowed to the University of Missouri by the federal government as part of the Agriculture College Act of 1862. The School of Forestry, Fisheries, and Wildlife took responsibility for the area in the early 1900s and, in 1946, began managing the area for research and as a forestry summer camp. The Conservation Department acquired the area in 1988 for the purposes of continuing forest research, creating forest products, woodland-related recreation opportunities for the public, and demonstrating the effects of forest management.

From 1947 to 2011, every forestry student spent one of their summers at the camp. It was a whole semester of learning crammed into those six short weeks.

For many of us, this was the first exposure to things like dozers used as skidders, log trucks, and sawmills. It was not only fun, but extremely rewarding to practice our tree measurement skills on standing trees, and to then cut them down, skid them out, load them on the log truck, haul them to the sawmill, and watch as they were milled into lumber. Then, we got to compare our estimates as the tree stood against the actual board feet that were cut and stacked.

Looking back, it was nice to create our own topo maps by doing the old school ways using a clinometer, compass, and cloth tape. These days, I have my own LiDAR scanner I can mount to a drone and fly and collect the data that can make a topo map of a property with centimeter accuracy in a fraction of the time, and with fewer ticks and poison ivy, I should mention.

Honestly, I don't remember a lot of the specifics of what we learned at camp, but I do remember the "community" we felt while there. There were 14 of us that summer. We dined together, worked together, lived together, studied together, played together, partied together, and just existed together. The staff who served us meals at the mess hall became family. Even the professors let us see a different side that you don't get to see during a normal semester.

Unfortunately, I lost track of most of the students who were there that summer. Of the 13 others, nine I lost track of as soon as I graduated. I have absolutely no idea what happened to them other than they scattered across the country. One would go on to be a leading dendrochronology expert and professor. One would lead an agency at the highest levels. One worked into upper management, and one seems like they are living the dream of still being a forester in the woods every day. I would like to think that all 14 of us have gone on to fulfill our own destinies, whether in forestry or not. Even the professors who were there that summer have all either long retired or passed on. So much knowledge is lost, unless those of us who were there can keep doing and now teaching the younger generations of foresters.

That was me learning how to skid logs out of the woods with a JD350 dozer. One of the other students gladly took the photo with my camera for me.

I then took a turn loading up the log truck. For many of us, this was the first exposure to logging operations. One of the other students gladly took the photo with my camera for me.

Chapter 5

My First Conservation Job

People always ask me how I was fortunate enough to get a job in conservation. The truth of the story is actually pretty crappy. Now, before I get to that part, let's back up a bit. Ok, let's back up a lot. Growing up outdoors, hunting, fishing, trapping, camping, and all other manner of being outside caused me to want to work outside. Add in the fact that my Uncle Rich was a Forester with the Corps of Engineers and fought wildfires as well had a huge impression on my younger self, and by the age of 12, I knew I wanted to be a Forester when I grew up. I didn't fully understand what that really meant at the time, but I just knew.

Fast forward a number of years, and I was looking for summer work. I put in applications all over the place, just about any place that dealt with trees or worked outside, tree services, nurseries, garden centers, and the Conservation Department. Most I never heard back from. A few called and did working interviews. I took a job working for a small tree nursery planting trees, but I was looking for more. Then I got a call from the Conservation Department from someone at Burr Oak Woods in Blue Springs. His name was Larry, and he invited me to come in for an interview.

I don't remember much of the interview, only a few key parts. I'm sure Larry asked me the normal questions like why did I want to work there and what my plans were for school. At some point, I mentioned that my uncle was a forester, and that's why I wanted to be a forester. He asked who my uncle was, and when I told him, Larry just sat back and chuckled. He knew Uncle Rich! Larry was the Farm Forester out of Clinton earlier in his career, and he and Rich had developed a friendship. They would go fishing together and argue about the pros and cons of live bait versus artificial lures.

Larry went on and on talking about days gone by and his old friend. I was just wondering if we would ever get back to the interview. Larry kept asking questions. But the questions were more fishing-related. Did I prefer live bait or lures? Where were some of Uncle Rich's favorite fishing spots? Did I like bank fishing or going out in the boat? Have I ever gone fishing in a canoe? I don't recall ever getting back to actual interview questions.

Larry wrapped up and got up to walk me out. No mention of the interview or the job, or anything. I was a bit bummed. As we got to the door, Larry opened it up, shook my hand, and thanked me for coming in. He then asked if I could start the following Monday. Well, yes, I could! Such a strange interview. On the drive home, it occurred to me that sometimes in life, it is who you know. I got the interview on my own, but I got my foot in the door because Rich was my uncle.

Truth be told, we talked fishing so much, I didn't even know what the job really was. I just knew I wanted to work for MDC as a forester, and I had better get some related experience. Monday comes, and I am there ready to take on the world of conservation work. We did the paperwork thing and got all that official stuff squared away. I was shown around the shop and the guys I would be working with all summer. I was dreaming of tagging along on timber sales and tree inventories. Then reality hit.

As a summer hourly, I was to be paired up with another summer hourly, and our job all summer consisted of just two things. First, we were to haul and spread wood chips across the miles and miles of trails spread out across the area. There was a huge chip pile at one of the parking lots. Larry allowed some of the tree services in the area to come dump their wood chips after they cut and chipped trees. This was a win-win. They got to dump for free, and we had an endless supply of wood chips for the trails. We had access to a UTV, a tractor with a loader, shovels, and rakes. Rain or shine, this was to be our main job that summer.

The second job was far less glamorous. It was reserved for Monday and Friday mornings. Only twice a week? It can't be that bad, right? First thing when we arrived on Monday, we were to grab the truck with a water unit in it and head around to the outhouses, or pit toilets, that were located at the trailhead parking lots. Yep, we had to clean and scrub the outhouses. Twice a week. Now you get why I said this was a crappy job.

I learned a lot about "people" that summer. How can anyone think that spreading their own crap all over the outhouse walls with their hands was a good idea? There were no sinks in the outhouses, so what was the upside? And yes, there was toilet paper left, so I don't think it was someone sitting there in a compromising situation. Yet, it seemed as though this happened almost on a weekly basis in at least one of the outhouses. We would do our best to spray them out. Grab the long handle brushes and scrub the walls, spray again. Squirt cleaner from some spray bottles and spray it all down once more. Refill the toilet paper bins. Move on to the next one. Every Monday morning. Every Friday morning. All summer long.

Now, this is not what I really thought conservation work was going to be, but I was smart enough to understand that conservation jobs were at a premium in those days. So, I worked hard and never complained. I may not have been the best ever cleaner of the crap house, but I acted like I was. Even when my co-worker started showing up late on Monday mornings, I still got the job done. And those trails? I was pretty proud of those trails. It is anyone's guess how many cubic yards of wood chips we spread that summer. I tried my best to keep the chips as neat and tidy as a trail through the middle of the forest would allow.

Funny, at the end of the summer, Larry asked if I would be interested in coming back the following summer. My co-worker decided to pursue other options. Larry even said my days of cleaning crappers were over. Sometimes in life, it is your work ethic that matters.

Fast-forward to my summer at Summer Camp? Yeah, that summer when the prospect of full-time employment was a slim chance? Larry brought me back on for those few remaining weeks between summer camp and our wedding, right before the semester started back up. This time, I was allowed to work on actual forestry projects like forest inventory and sick tree calls with other forestry staff in the District. Some real experience and a few bucks made to salvage the summer budget.

The following May, I graduated from Mizzou, and with Tammy having an internship in the Kansas City area, I again asked Larry for a job. He welcomed me back, and even though I was hourly, he treated me like a full-time employee. I not only went with the other foresters, but I was even given projects to complete on my own. I had heard a rumor that there was a potential full-time permanent ARF job coming up. ARF stands for Assistant Resource Forester and is the entry-level forester job with MDC, although years later, they changed it to Resource Forester Assistant. I think it had to do with someone being offended that their "job title" sounded like a dog barking or something. I honestly still refer to that position as an ARF. I think it adds to the story of someone earning their way.

This goes hand in hand with another term Larry had. He referred to new foresters as PFWs. He always said that and then laughed in the way that only Larry can. One day, I finally asked what PFW stood for, and he explained it was a term for new foresters that came from out west and stood for Piss Fir Willys. Subalpine fir is many times also called piss fir due to "no one giving a piss about it," and therefore that's how seasoned crews felt about the new crew members or new foresters. It is akin to FNG in today's vernacular, you know, Freaking New Guy, or something like that, I'm told.

That is another thing I appreciated about Larry. He didn't cuss. I mean, piss and crap were the absolute limit of what I ever heard come out of his mouth, and even then, it was with an almost regretful embarrassment or maybe a giggle.

Anyway, back to the potential permanent position. Larry started coaching me on how the interview would be, what to work on with responses, and how to handle some strong personalities that would surely be on the interview panel. The interview came and went. It was grueling, as I remember. In between each set of questions and answers, I was reminded how improbable a full-time job was with MDC, how many qualified people had applied for this single position, so on and so forth.

I went back home, hoping for the best, preparing for the worst, but ready to take what was to come. Thankfully, I was selected for the position and started my true full-time career as a forester barely two months after graduating.

So, back to the original question of when someone asks how did I get into being a forester? I say it is a combination of many things. Yes, who you know can get you in the door, but it also takes what you know and a good work ethic as well. It means taking the job that no one else wants to do, and doing that crappy job as well as you can, and accepting minimum wage and being grateful for all of it. Take pride in your work. Network and use your resources. That is what is needed to get you into a job you want. Then, from there? Don't stop. You need to earn your right to keep that job every day with the same work ethic and pride in a job well done. Constantly learn how to do the job better and keep learning.

This is what can lead to a fulfilling, long-lasting career. And I am so thankful for those early influences and experiences that shaped me from the very beginning.

My official first day as a full time employee.
Photo by Tammy Shroyer

Chapter 6

Peak Fall Color

I think everyone enjoys the colors of fall. It is yet another display by Mother Nature to show off, and she sure shows off in grand style, right before the gray of winter sets in. Kind of like a capstone to the year, if you will. Of course, there is a segment of the population that doesn't like fall. They may just see the warmth of summer about to leave, or lament the vast amounts of leaves that they are soon to have to rake up in their yard. I suspect, however, that even with all that negativity, they still, secretly, enjoy the colors that greet their eyes every day.

For me, fall is definitely my favorite season. In addition to the colors, I enjoy the slight coolness in the breeze at the beginning of fall, signaling the coming winter cold. I look forward to throwing on that comfortable hooded sweatshirt or layering up in some merino wool. Fall is just a wonderful time to be in the forest, whether it's for work or walking a trail with the family. If you check out any random advertisement for a vacation destination, or a community trying to get more people to visit, or even move to their area, there is a good chance that the first photo you will see will be taken in the fall with the trees in full reds, oranges, and yellows.

With all this happening each year, it is only natural for some folks to want to capitalize on that color. I have heard that some retirements are put off until they can retire in the fall, as they may relate to the symbolism of the fall time of their life, or high school students want their Senior Photos in the fall. Those who prefer a fall wedding want the colors to be (almost) as vibrant as their wedding dress is white. If a community has a fall festival, it wants the peak of fall color.

What causes fall color anyway? In simple terms, the cooler and shorter days signal the tree that winter is coming, and it needs to start preparing. As less chlorophyll is made, and starts to break down, other pigments that are in the leaf can then begin to show through. There are a lot of smaller variables that go into how the color will be in any particular year, but I won't go into those here and bore you.

As my first work location was in the Kansas City District, I was officed at Burr Oak Woods. All of KC Forestry worked there, except for one of the Urban Foresters. Now, when I got into forestry, I figured that cubicle work was never in my future. However, that grim reality rudely slapped me across the face. What should have been an office for one, maybe two people, had dividers and typical cubicle work stations installed, and there were four of us crammed into that one space. The four spaces were for me, as the newly minted ARF, the Assistant District Forester, the Forest Management Technician, and the other Urban Forester. With everyone in such close proximity, phone conversations were very public. Remember, this was before cell phones, and even email hadn't caught on all that well. Most of our business was conducted by phone or in person.

Jerry, the Urban Forester, took a lot of calls all year round, but it seemed to pick up the pace in late summer and early fall. I overheard him take a few calls one day and, in each call, he related the date of October 23rd. After hearing this for the third or fourth call, I asked him what was happening on October 23rd? Did I miss something that should have been on my calendar? He stated that the peak of fall color this year was going to be October 23rd. Oh, ok. Thanks. And I went on about my day.

As more days passed, there seemed to be more calls asking about fall color. Naturally, some of the calls started hitting my desk when I was in the office. Parents asking about fall color for their students' Senior Photos, a last-minute Fun Run being scheduled, those sorts of things. I fumbled through trying to give the scientific explanation of why we can't really predict the peak of fall color and all the subtle variables. Most calls ended up with both me and the caller frustrated. Jerry, listening to all this, just chuckled each time.

Ok, fine. I give up. The very next call that came to me asking when the peak of fall color was going to be, I just matter-of-factly stated October 23rd. Ok, thanks! Click. I hung up the phone, leaned back in my chair, and must have said something like, "Wow, that was easy." At this point, Jerry gets out of his chair and comes around the divider and, with a big smile on his face, says, "The young one is finally starting to learn."

He then proceeds to explain to me that although I am correct in my scientific explanation, most people don't care or even want to try to understand. They just want a date. In addition, you can't even look at any particular day and with any confidence say that last Tuesday was the peak. However, we kind of know what week or so was the peak, and his 30 years of doing this, that week seemed to average out to about the first part of the 4th week of October. He just refined it and started saying the 23rd. He found that the phone conversations became much less lengthy, as I had just experienced. We both laughed a bit at this new lesson I had just learned, and we went about our day.

A couple of days later, Jerry hollers over the divider, "Hey, call for you." Nothing more. He just transfers it to my desk, and the phone rings. I answer. It is a young lady who is trying to plan a wedding for the coming weekend or the following weekend and really NEEDS to know what day the fall color will be the best. Without missing a beat, I state October 23rd. "But the 23rd is on a Wednesday, I can't get married on a Wednesday." Sorry, Miss, the 23rd is when the peak will be. "Oh, well, ok, thanks. I'll see what I can do." With a short congratulations on her upcoming wedding, I say bye and hang up. Clapping immediately started from all three other cubicles, and Jerry exclaimed, "I think our young forester has just graduated!"

And so began the rest of my career as it relates to the peak of fall color. A lesson learned just three months into my first full-time job has stayed with me to this day. Anytime anyone asks me about fall color, my answer is the same: October 23rd. I say it with confidence, with conviction. Very seldom is it even questioned, and if so, there is a little resigned fact that they may have to adjust plans. Overall, I must have nailed it every year, as I have never had anyone call back and say I was wrong or missed it by a day or two.

My family has been in on this joke for decades now. People in the community know my business is trees, so it seems I can't make it through the "Walmarts" during the fall season without getting asked the fall color question. Therefore, the number 23 has become kind of a default answer for a lot of things around our family. It is just a known and accepted answer, on the same level as 42 is to life, the universe, and everything. Thanks for that important lesson early on, Jerry, very much appreciated!

Judging by the fall color, we must be pretty close to October 23rd.

Chapter 7

Don't Fall Asleep

Working with our Forest Management Technician, Terry, was always an adventure. He had worked at a number of locations around the state and was at the tail end of his long and storied career. I mostly worked with Terry on screening excess property or doing fire department visits. Occasionally, we would collaborate on a forest management project on one of the state land areas.

Terry was always good for a story. I guess after a full career, you have a lot to choose from, and I don't know that I ever heard the same story twice. He was also a bit of a storyteller, which is why I think I liked him so much. A little embellishment here, a stretch there, but always based in truth. He also had a little bit of a practical joker running through his blood. Terry was always smiling, and I don't know that I recall ever seeing him get mad. Maybe I'm just suppressing those memories, who knows?

Terry had one rule. Don't fall asleep in the passenger seat while he is driving. If he had to stay awake to drive, his passengers had to stay awake as well. No free rides, and this was not a taxi service. Like I said, I don't recall seeing him ever get mad, but I did see him get real serious one time. We had another employee from a different division with us one day. This employee, who will remain nameless, twisted around and got real comfortable in the front passenger seat one afternoon on the way back to the office. Terry stated his rule, much to the dismay of said individual. They spouted back about a safety nap or something like that, and Terry's face went cold and expressionless. He very matter-of-factly reminded everyone in the vehicle that we were still on the clock. There was no sleeping during work hours. Nowhere in the policy did it say that sleeping while on duty was allowed. On and on he went. Needless to say, the passenger didn't close their eyes, not that they would have been able to sleep with Terry droning on like he was.

After we got back and were finishing up in the shop bay for the day, Terry was laughing and grinning, recounting the afternoon ride. "Did you see so and so when I said this?" At this point, I was starting to wonder if this rule of his was hard and fast, or just a ruse, or just a way to not have to drive alone in silence. Joking or serious? I don't know, to be honest.

Well, fast forward a couple of months, and this lesson is always at the front of my mind anytime I am riding with Terry. To his credit, he never slept while he was a passenger, either. Come to think of it, maybe he was just terrified of a young new forester and didn't trust my driving. The day finally came when I was struggling to keep my eyes open.

We had a long day in an area a ways East of the office. We fixed and repainted signs, dug a hole for a new sign, and moved a bunch of rolls of wire for a future project. Nothing out of the ordinary, really, but it was a hot day and we were way over on our hours and getting back late. So it was late afternoon, the sun was getting lower and lower in the sky, and we were driving straight into it. Of course, being a humid day, the haze made the sun even more unbearable. My eyes were just screaming to close, and my brain just wanted a brief nap. I tried to keep chatting to stay awake, but Terry just wasn't talking back. It was like he wanted me to fall asleep.

I was apologizing for being so sleepy and making excuses. I really couldn't read Terry very well at this point. Was he mad at me? Was he just as tired as I, but trying to drive? I think I was even slurring words at this point. My mind and body finally give up, and I fall into a peaceful, restful sleep....

"Aaaaahhhhh!!!!" Terry was screaming at the top of his lungs as I got jolted awake and felt the jeep swerve. Terry slammed on the brakes, and tires screeched, and he never stopped screaming. I'm awake! I'M AWAKE! Every nerve in my body was awake. I think even some nerves from some of my ancestors were awake at that point, too. I looked around to see what was the cause of all this mayhem, blood pumping 90 miles an hour. Nothing. We are stopped in the middle of the drive and about to turn into the shop parking lot.

At this point, the yelling has transitioned into a full, rolling, uncontrollable laughter. I yell, "What was that all about?" I was pissed. Terry got control of his fit of hysteria and calmed down enough to smile and say, "I told you never to fall asleep while I am driving, now you know why." Dang, he got me good. Apparently, he was trying to get me to doze off, and I played right into his plan. I hadn't realized it, but I dozed off about 10 minutes from the shop, and he was just waiting until we were back in the area to implement his devious plan. We weren't going very fast, and no one was around, as it was after hours. He knew he was going just fast enough to lock the tires up and threw in a good swerve for full effect.

I was none too happy about any of this, but I did kind of see his side of things and eventually came to laugh about it. Eventually. I will say, I never did doze off again while Terry was driving, although I was able to witness it once with another unsuspecting employee. I even contributed by screaming when it was time. I also learned that he would not ever do this on the highway or anything dangerous.

Terry and I would go on to share many other adventures in the few years I was at Burr Oak Woods. He was definitely from the "old school" of the agency, and I appreciated all the stories, friendship, and mentoring he provided. And, as with most lessons learned early in my career, this one stuck with me. Even today, if I am riding along with someone, I think back to that day as I close my eyes for a quick nap.

Illustration by Paige Schooner

Chapter 8

Busted Ankle

The very nature of being a forester means a lot of days working alone. I am rather fond of working alone, actually. I also used to have a bad habit of not always telling someone where I was going for the day. While I was still being trained and spent many days with someone from the District, specific to their expertise, I was also given projects to work on by myself. This was part of the process. See how I can manage my time, balance multiple tasks, and compete with priorities.

The Kansas City District was not all that large, and you could get to most work locations within an hour or an hour and a half. Anyway, one day I decided I needed to head to a private landowner project over in rural Ray County. This was a few hundred acres, and I needed to look at the potential for a timber sale on the back side of the property. This was an established, long-term cooperator who lived out of state, so there would be no problem just showing up and doing what I needed to do.

It was a rather beautiful day, as I recall, and I knew I had a long day ahead of me. I parked at the gate and figured that instead of walking the trail around the long way, I could save some time by cutting straight to where I needed to be. I mean, as a forester, that's what we do, we walk everywhere, and through the forest is part of the job. Besides, walking the trail was boring. I needed to end up about a mile from the road, and about three-quarters of the way there, I would have to cross a creek. No worries, it had been dry lately, so I knew I would be able to find an easy spot to cross.

When I reached the creek, it was a beautiful setting. I had just come down a fairly steep slope to the creek, and there was a broad flat on the other side with an open grassy field that I would need to cross the final couple of hundred yards to reach the stand I was after. Due to getting out of the office a bit later than expected and the drive, I decided to stop here and eat an early lunch. I grabbed my sandwich and a few chips from my cruiser vest, found a large rock, and ate my lunch. I surveyed up and down the creek to plan my crossing. The water was down, and, in most places, it looked like you could easily jump across and not get your boots wet. The only real issues were the wet algae and moss on the rocks. These made things look a little dicey.

After finishing my sandwich and chips, I washed it all down with a can of Mountain Dew. Yeah, I know, not the best form of hydration, but I was young and energetic and had grown up on a breakfast of Mountain Dew and honeybuns every day for years. I was the model of healthy eating! With all the trash put back in my vest, I headed to the chosen spot to cross. Crossing the creek was fairly easy, a couple of long strides to some dry rocks, and I was past the water. Five more feet of rocks to climb out of the creek bottom to the flat, and I was in the clear.

Except, I wasn't. The last step on the rocks, the rock rolled. My ankle rolled with it, and I lost my balance. My foot rolled into the space between 2 other rocks that didn't feel like rolling, and my body was headed the other direction. Ouch! I don't recall hearing anything pop, other than the rocks hitting each other, but I was in serious pain. I got my foot twisted free from being caught by the bigger, non-movable rocks, but I knew the damage was done. I couldn't put any weight on my ankle. The pain was intense, and I could sense the swelling starting. I didn't take my boot off to look, but I did re-tighten my laces and really cranked down on my ankle to try to keep my ankle from swelling like a grapefruit. At this point, I started to survey my situation.

Now, this was not the first time I have been presented with a situation that seemed to have few options. I had learned through growing up to be self-sufficient (think Gen X kid), had learned a number of critical thinking skills, as well as backcountry skills from being in Boy Scouts, and earning my Eagle Scout. This problem, like many, just needed to be divided into the liabilities presented and the assets available to you.

Let's start with the liabilities. There I was, with a busted ankle that I can't put weight on. I'm on the far side of a rocky creek bed with wet, slimy rocks. There is a steep slope just on the other side of the creek, and it was about three-quarters of a mile to the truck. No one else was on the property. No one knew where I was for the day. No one would know where to even begin looking for me. And just in case you were wondering, this was the dark days before cell phones. Finally, I have no water to drink, at least not clean water, as I am right next to a creek.

I like to try to look at the positive side of things, so now I could look at the assets I had. I just finished lunch, so I had that boost of calories, and it would be a while before hunger started to annoy me. I was jacked up on caffeine, so I was feeling like I had energy to commit to solving the problem. Yes, when the sugar and caffeine wear off, this will probably slide over to the liability side of things, but for now, I'm good. I know that I can remain calm in these situations. I know I am in pretty good shape. It is still before noon, so there is a lot of daylight remaining. Finally, I am wearing my bright orange cruiser vest.

I knew that self-rescue was the order of the day. No one was coming to look for me, as no one knew where I was. First things first, get back across the creek. Using a three-legged approach, both hands and one foot, I sort of bear crawl over the rocks to the water's edge. No more worrying about staying dry. Luckily, the day, and if I ran into an overnight, temperatures were going to be manageable. The rocks were slick and slimy out of the water, but underwater, they weren't so bad, so I had a better grip with my hands. I let my busted ankle sit in the water, more to be sure both boots were soaked, as I think that would have driven me crazy to have one wet and one dry foot. The cool water actually felt good on the swollen ankle. After what seemed like forever, I made it all the way across the creek bed and to the slope of the hill, where I could use some trees to help me stand and balance.

At this point, I am lamenting the fact that my list of assets does not contain a saw of some sort. I could have cut a small tree and fashioned a crutch to help. However, all I had were some pieces of driftwood from the creek, and they were not as strong as live wood, but they would have to do. Against my better judgment of trying to stay in an open area where I could be seen, I did veer towards the thickest part of the hill, as I would have a lot more small trees and bushes to grab hold of to help pull myself up the hill. The risk paid off, and I was able to grab and pull my way up the worst part of the hill to flatter ground. I don't know that I ever looked at my watch to see what time it was. Maybe I didn't want to know, the fear of time getting away from me. It really didn't matter. Time was irrelevant at the moment, but I am guessing the self-rescue process was many hours in at this point.

Once on flatter ground, I was finally able to hop a few hops at a time and rest against a small tree. Every now and then, I tested the ankle. The pain was intense, but either I was miraculously healing, or I started to curse the pain away, and I was able to put a pound or two of pressure on that leg. This was a huge mental boost and gave me renewed energy. Hop, hop, step, hop, hop started to become the rhythm. Actually, it was more of a painful hop, painful hop, extremely painful step, painful hop, painful hop, long rest up against a tree, catching my breath.

The day was starting to get towards late evening when I was finally within sight of the truck. Again, a renewed energy. I was soaked through with sweat, extremely thirsty, exhausted, and still in a lot of pain, but I could now see my desired destination. With a little extra break to build up my energy and confidence, I took off on the last leg (pun intended) and stopped to rest fewer times than before. I just gritted my teeth harder through the pain but focused on the finish line.

I got to the truck right at sundown, and it felt so good to sit and rest in a somewhat comfortable position. Thankfully, I had a water jug in the truck. Now, you might think my ordeal was over, right? Wrong. My truck was a standard transmission. Three pedals to operate with two good legs is a no-brainer, but three pedals with only one functional foot? Yeah, that took some trial and error, but I was finally able to figure it out. I didn't go to the office to swap vehicles that night. I went straight home.

For the most part, I have learned to tell someone where I am going and when I plan to return, typical travel path, things like that. At the office, we started an In/Out Board. I say "we" as if I were part of that good idea, but in reality, it was only after an earful from my boss, Larry, and an even more severe earful from the real boss (read Wife), that I realized I was the reason for it. In a few years, I got my first bag phone to keep in the vehicle.

Nowadays, with cell phones, it is so much easier to communicate. However, there are still many places with less-than-ideal cell coverage, so it is always a good idea to tell someone where you will be and when you plan to return. Some of the lessons I learned early on were painful ones. But I did learn from them.

Illustration by Paige Schooner

Chapter 9

Flooded Inventory on the Platte

During my time on the Kansas City District, I was tasked with getting some area compartment forest inventories completed. This one particular year, 1998, I believe, I was working on the Platte Falls Conservation Area inventory. As is typical on just about any public land in Missouri, you really can't get much work done until after the hunting seasons, so work got started in mid-January. I made slow but steady progress through the winter months. Short days, a drive from the office, other projects and priorities, and helping others on their projects all contributed to the slow part of that project.

Most of the inventory was without issues or concerns. In fact, some areas were rather cool to be in. For anyone who is familiar with the area, you know the defining characteristic of the river is the "duck head." This area is a broad, flat, bottomland forest that is defined where the river runs very close to itself, less than 250 yards. This is what is called the "neck," and the river runs over 2.5 miles around and in the shape of a cartoonish duck head, encompassing about 160 acres. Working in this area provided a couple of days of nice solitude amongst some rather large trees. I remember a rather large rookery of some Great Blue Herons in a patch of large-diameter sycamore.

Most of the area is in the Platte River flood plain, but there are a few areas of steep upland side slopes. As the late winter and early spring were getting pretty wet with rain, I started saving some of these acres for when the river came up and flooded the lower stands of forest. This worked well for a while. However, about late April, I was out of the upland forest to inventory and down to a small handful of stands back in the flat bottomlands. Unfortunately, the river was up and continued to stay up with no real relief in sight. In addition, I was running out of time as the project needed to be wrapped up in a couple of weeks.

As much as I tried to put it off, I knew what was coming. I monitored the water levels daily, and finally, the water dropped enough that I could get into the final stands to complete my inventory. Being a duck hunter, I had my own chest waders. So, there I was, three stands left to complete. An easy half day in normal conditions. But, wading around in flooded bottomland forest in anywhere from one to three feet of water was not a normal inventory day. To put it into context, I do enjoy hunting flooded trees for mallards in the late season, but that is usually on a cold winter day, where we walk or wade into a spot and then stand there with only occasionally retrieving a duck now and then. Forest inventory in flooded trees in early May was a different story.

The water was muddy, and the stands I was working in were full of smaller-diameter, younger trees that caught every piece of drift that came by. The ground and footing were uneven, and you could not see through the muddy water to plan your steps. It was warm, or should I say hot, especially in waders designed for winter hunting. I think I may have been just as wet from sweat as if I didn't have waders at all. I will admit, I was a little uneasy. This was so far out of my comfort zone and was such a unique experience. I know at this point, any foresters from down south are probably saying, "just another day, at least you don't have gators," but it still bothered me.

Funny, cold water while hunting was not a problem. Warm water with a lot more walking and ground to cover was a different story.

Ultimately, what should have been a 4-hour project took me three different days that week to finish up. The area had been flooded for so long at this point that the smell was getting rank, not to mention my sweaty waders! Later in my career, there would be other opportunities for flooded forest inventory and even flooded timber sale marking, but none would have the same impact as that first time. Yes, I was very glad to be done with that project. As a young forester who was having lots of firsts, this ranked right up there.

High water can't stop work, only hinder it. Thankfully this log went right out to the tree I needed to get to. Photo by Tammy Shroyer

Thankfully the water had receded on this project to only ankle deep in places. Plot center stick and flagging.

Chapter 10

Blown Tires on the Jeep

This story ties back to the fellow employee dynamic we had at the Kansas City District office. Remember, there were 4 of us in the one office, and Larry, the District Forester, was in his own office. Larry was a great boss as far as I was concerned, but of course, I got to hear the other stories from the guys. Stories of penny pinching, tight budgets, squawking over any extra money being spent, that sort of thing. I, of course, had not seen this side of Larry yet and was not really looking to find out any time soon.

One particular day, I was only going to work a half day as I needed to take the afternoon off for some personal errands, and I was to meet Tammy no later than 12:15! Due to this half day and the desire not to be in the office, I decided I would check on a tree planting in the far North part of Burr Oak Woods. The trail was a little old and overgrown, so I asked to borrow the Jeep instead of taking my full-sized Dodge. I figured the Jeep would weave around the overgrown spots a bit better, so off I went with not a worry in the world.

The service road I was looking for was North of Pink Hill Road, and as I turned off the pavement, I noticed the service road was a bit more overgrown than I expected. I also noticed that the old 2-track was a bit eroded due to some heavy rains we had recently experienced. Oh well, I've seen all the Jeep commercials, these would be fun and no problem for this beast. I slowly crawled over and around rocks, washouts, small branches, and whatever else had fallen into the roadway that seemed long abandoned. I mean, when was the last time this road was mowed or maintained? It should have been yearly, but it appeared that the last time this was used was by some wagon from some settler in about 1820!

I just needed to make it a little over a half mile along this service road to get to where I wanted to be, but it seemed like forever to navigate the washed-out, unmaintained road. I finally made it to my destination and did what I had come to do. As the time was getting away from me, I figured I had better start heading back.

Maybe it was the reversed perspective of heading the other direction on the road, but it appeared to be better maintained, so I figured I could drive a bit faster than I had come in. In no way was my hurry affected by the fact that I was on the verge of running late. No way at all. The bumps were a little bouncier, and the washed-out tracks zigged and zagged. This was fine, except that when I wanted to zig, the Jeep zagged, and then when I thought we were going to zag, we zigged. It was during one of these uncoordinated zigs when I wanted to zag moments that I heard a loud "pop" immediately followed by a loud hissing sound. Funny, I don't recall the Jeep commercials saying their vehicles made these noises.

As I was only about 75 yards from the pavement, I figured I could make it down what was left of this little hill. I figured wrong. Both tires on the driver's side were completely flat. Add in that I was at one of the deeper washouts, and you get a high-centered vehicle that refuses to budge. Luckily, I was able to get my door open, but just barely. Taking a look around, I used my highly developed detective skills to describe the scene.

At the spot where I had wanted to zag, the vehicle decided, much to my dismay, to zig and threw the front tire into a sharp, flat rock that was protruding from the aforementioned washout. This produced the pop sound I had heard, and as the momentum of the vehicle continued to move us forward, the rear tire also hit the same rock. However, I think the rock felt sorry for causing so much trouble, so it sliced the tire just enough to break through the sidewall. Yep, that was the hissing sound I heard. My detective skills were firing on all cylinders!

Well, this was a predicament. At this point, I am really in danger of running late. The only thing I could do was call for help. As even I know basic math, I know that one spare tire is not a fix for two flat tires. I called Terry on the radio and asked if he could meet me at the service road location. I was not about to say over the radio what was going on. Soon enough, Terry arrived. He jumped out of his truck and just started laughing and saying, "Oh, you are going to get it now! Larry will be all over you! Two blown tires, these can't be repaired." Yeah, I know, thanks for stating the obvious.

I still have a problem. I needed to be gone 20 minutes ago, and the Jeep is still high-centered with two flat tires. This is also where I will say that I worked with some great people. Even though you don't fall asleep while riding with Terry, he has your back no matter what. He told me to take his truck and head back to the shop, and when I get there, tell the others what had happened and for them to bring the tractor and another truck back out, and for me to get out of there as quietly as possible.

I did just that. Of course, I took some ribbing from the shop guys, but I was able to get in and out of the office without being seen and headed home. This being before cell phones, I had some explaining to do when I got there as to why I was late, but we were fine, nonetheless, on our afternoon endeavors.

To this day, I really don't know what transpired. They never told me, and I can only assume what happened. See, we had a tire repair machine in the shop for normal flats from nails and such, but these tires were ruined with slashes in the sidewalls. New tires would have to be purchased, and this purchase would be scrutinized, and questions would be asked.

When I returned to work the next day, there was no mention from anyone about what had happened. I kept waiting for the hammer to drop and get called into Larry's office. Days went by, then weeks, then months. Nothing. Even when Terry and I were alone on projects, it was never talked about. Some may surmise that Larry never heard about it, and the purchase was somehow kept quiet. I, however, have a different idea of what happened.

You see, Larry really did know everything that happened in his District. Nothing got by him. But I also think something else was at work here. I ended up being the last Assistant Resource Forester that Larry ever hired and trained up, and I believe he has a soft spot for me. Although Larry and all those in the District at the time will probably deny it, I believe this is the absolute truth. Larry had to know what happened that day and had to approve the expenditure, and to everyone's amazement, he just chose not to make an issue of it.

Even now, almost 30 years later, I dare not ask Larry about it. I don't want anything to change my perception of being his favorite forester! Oh, and Larry, if you are reading this, please forget you ever read this story!

Illustration by Paige Schooner

Chapter 11

Hole In the Wall Eateries and Peanut Butter Pie

My time spent on the Kansas City District with Larry as my boss during my formative years was really quite something. Ha, in fact, that part of the previous sentence came from Larry. He was fond of saying, "Isn't that something?" when given some interesting information. I guess I have picked that up from him. Anyway, as a young forester, I was eager to pick up anything and everything from those who would take the time to teach me, and I really enjoyed the days when I got to go with Larry.

I picked up on a few peculiar habits of Larry's early on. No matter where we went in the District, we never ate at a known restaurant. Now, by that I mean, no fast food, and none of the bigger chain-type restaurants. We always ended up at some hole-in-the-wall type place. Now, being a small-town kid from Wellington, I sort of understood this as one of my favorite places to eat growing up was Brown's Mexican just off the square. Everyone in town knew the place, and it was good; no, it was great food. But to the casual passerby, if you looked at it from the outside, you might not think too much about it.

On the other side of my coin, so to speak, as a kid from a small town who now understands what introvert and extrovert mean, I was not one to try new things. I liked my routine and didn't like interacting with new people if I didn't have to. But Larry, on the other hand, didn't know a stranger. Maybe it was truly his nature, or maybe he had just been around so long that he had met everyone already. Regardless, we stopped at some rather interesting places for lunch.

Some of these places you will laugh at me when I mention them, but you have to remember, I was new and learning new skills and getting "cultured," as Larry put it once. We can start with Arthur Bryant's BBQ. Anyone who is from KC, visits KC, or has even heard of KC knows about Arthur Bryant's. I think Larry dreamed up excuses to be in that part of Kansas City close to lunch. Yes, it is now one of my favorite places to go when I get back that way.

Sticking with BBQ, there was Bates City BBQ. I was already familiar with Bates City BBQ, and the greasy fries were actually a favorite of mine. Yet another place I try to make time for when I travel back through the area.

There were other places that I can remember, like the 10-13 Diner in Richmond, named for the intersection of Highway 10 and Highway 13. If anyone remembers the 10 codes, 10-13 was generally used for weather updates or conditions, and in the case of the restaurant, the conditions meant good food was just an order away. Another spot Larry introduced me to was just up the road in Polo at the Red Rooster Cafe. Even though this was technically out of the District, any time we were in northern Ray County, we seemed to make the trip to Polo for lunch.

Those that I have mentioned so far ended up being places I went back to repeatedly, and although they may not be true hole-in-the-wall types, these are definitely places Larry liked to go and exposed me to some good food. However, there were a number of other, long-forgotten places and names that Larry dragged me to that I never would have gone on my own. In fact, I didn't even want to go at the time. They just looked shady. Some didn't even look like they were going to serve food, but that didn't deter Larry. In we went, and once inside, there was indeed a restaurant. Larry greeted the waitress like they were old buddies. The food was always better than I expected, which may not say a lot, as I didn't really have high expectations, but that was on me.

I should probably mention at this point that Larry always, and I mean always, asked for peanut butter pie for dessert. Whether they had it or not, he always asked. The few times someone actually had it, I passed, as I did not want to try yet another something new. I had never had it and didn't realize what I was missing. I was just too afraid to try something new.

I never really thought about this early influence until recently. As I look back over my career, I can see glimpses of how Larry shaped me. I still like my routine and have been accused of being boring with my eating out and the places I choose. But, on occasion, I do feel a bit daring, and especially when I'm by myself on some consulting job, I will pass by the McDonald's or Casey's Pizza and look for a small, hometown Mexican restaurant or burger diner.

No, not every spot has worked out as the best food I have had, but some are on my list to go to if I ever return through some of those areas. A few examples include Los Mariachis in Anderson, El Mariachi in Cassville, Rich's Famous Burgers in St. James, and Hick's Hometown Drive-In in Chillicothe.

Others are on the list to hit every time I visit, such as Pizza Glen in Clinton. They opened in 1976, so I say I have eaten there literally the entire time of their existence. In fact, when we would have staff in for a meeting or training in Clinton, we always ended up at "The Glen" for pizza and a salad. Next, for anyone who may have gone through scouts in the Kansas City area knows of the little store in Iconium, and with spending most of my career in the area, and now working on the Scout Camp in my Consulting career, I always stop by for a rather plain bologna and cheese sandwich with mayo on white bread and a Peach Nehi.

Sadly, at the time I am writing this, they suffered a fire, and the store burned down, but they are rebuilding, and I am looking forward to helping them recover by stopping in once they are open again for a meal.

Yet another place I have to hit every time I can is the Maid-Rite Drive-In in Lexington. Yes, this is actually a chain based out of Iowa, but there are only two or three left in Missouri. Growing up, this was a favorite treat of my family, actually my dad's favorite, so naturally, it became one of my favorites. To casually drive by, you may not understand the importance of this place. Hidden at the bottom of the hill, not on the normal travel path of anyone but locals, sits the unassuming gem of loose meat burgers. If you haven't had a Maid-Rite sandwich, you really need to search one out. It is definitely made right!

These days, I do eat the occasional piece of peanut butter pie. I wonder why I didn't try a piece earlier. It doesn't matter to me whether it is from a small hometown restaurant or maybe from some yet unknown world-famous pie maker who only makes it for the local community or church get together, when I see a piece of peanut butter pie, I will quietly take a piece. I cherish each bite and let my thoughts take me back to the early days of my forestry career, think of Larry, and wonder what he may be up to. I can hear him now, knowing I am eating a piece of pie that I always turned down, "Isn't that something!?"

A sandwich and Peach Nehi Float at Scott's in Iconium.

Chapter 12

Eerie Days in the Forest

This topic spans my entire career, from a potentially haunted portion of the Bethany Falls Trail at Burr Oak Woods in my early days, to freak wind events on a rather calm day in the western mountains, to weather systems that caused historic wildfires on one side and devastating tornadoes on the other. Now, you would think that with all the time I had spent in the woods growing up, running trap lines, hunting, hiking, camping, and such, I would have seen or heard a tree or large branch fall in the forest. The first time was actually after I had been a full-time forester for about a year.

I don't remember why I was there, but for some reason or another, I was on the Bethany Falls Trail in the Northwest portion of Burr Oak Woods. I had been on this trail many times over the years, working hourly, chipping the trail, performing trail maintenance, or hiking after work or on a weekend. This time, like so many before, was a pretty nondescript day. A very light to almost non-existent breeze and a mostly high overcast that muted the shadows. Regardless, I don't remember much leading up to the event, as this was just another day. However, that particular day would soon become another first for me.

As I was heading west, just about to get to the large outcropping of rocks that the trail got its name from, Bethany Falls Limestone, a rather large limb came out of a tree about 20 yards to my right. No warning. No wind. Nothing. Just a sudden whooshing, cracking, and snapping as the limb came crashing down from about 40 feet high. I froze! My heart was pounding in my chest. This truly scared me and caught me off guard. I stood there for what seemed like an eternity, taking it all in, calming myself down. From where I was, I could see no apparent reason for the limb to break.

Unnerved a bit, I started on my way, and no more than 10 steps into my journey, a whole tree came crashing down. This one was about 50 yards ahead of me and slightly to my right. That did it. I turned and ran the short distance back the way I came until I came out at the open field. There, with no trees around me, I paused and took stock of my situation. What had just happened? I had heard of things like this happening, but never experienced it firsthand. Looking back, I should have gone and investigated. Was the limb break from a poor branch angle? Was the whole tree failing from root rot? All I know is that day, I decided to change plans and head back to the office. That was just a little too spooky for me at the time.

I asked around after that. Did the others at the office experience sudden tree failures in the forest for no apparent reason? We talked about their experiences, where and when they happened, and the causes. It was also brought up that some considered that part of the Bethany Falls Trail haunted. One story was that there was the spirit of a homesteader who still wandered that area. Whether that portion of the trail is haunted or not, I really don't know. I have always been cautious and paid extra attention when I walked that trail again. Nothing else has ever happened there, so I guess we can just chalk it up to a first of many such occurrences over my career, and firsts are usually more impactful.

As my career runs through the years and decades, I have accumulated many more interesting and downright eerie situations. During many wildfires, which usually seem to be on windy days, I have seen and heard many trees fall and branches break. One such event was predicted hours in advance. I was a Felling Boss on a fire in Idaho, and we knew some high winds were expected later in the afternoon. As we got closer to the time, there was an updated weather notification on the radio. The wind event was coming sooner and more severely than originally predicted. We got everyone out into open areas away from trees and hunkered down.

You could hear the wind front coming. While it was rather still where I was hunkering behind a large rock in the middle of a meadow, the sound from the south grew steadily louder until the wind hit. It seemed like we went from no wind to 70 or 80 or even 90 mph winds in the blink of an eye. What was weird was that it only lasted about 20 seconds and then completely shut off, back to still. During those 20 seconds, however, I lost count of the number of trees that came down. Most were whole trees that were uprooted. Many others broke and snapped about 10 to 20 feet up the trunk.

The suddenness of the wind's dying was what was eerie. The wind event had been predicted for hours, but with it only lasting 20 or so seconds and moving on, we were all a little on edge. We did radio checks and everyone was ok, but there were so many trees down across the roads that it took the rest of the day to get the roads cleared and get everyone back to camp.

Another eerie day had nothing to do with wind or trees falling. Again, just a normal day in the forest, working alone. Now, when I'm alone in the forest, I am usually pretty quiet except for the crunch of the leaves beneath my boots. This allows for the seeing of deer and turkeys and squirrels, and all other wildlife.

You can hear birds chirping, or hawks screeching, or the rustle of leaves in the wind. On this particular day, it seemed as though everything stopped. No sound at all. Even my own steps seemed like they were on mute. No wind, no birds, nothing. To say the silence was deafening was an understatement. I stopped. The only thing I could hear was my heart beating in my chest.

I am pretty comfortable being alone in the woods, but this was different. Was there something out there? What caused this sudden stop to everything? I studied everything I could see. Nothing was out of the ordinary other than the maddening quiet. I don't know how long I stood there, but eventually the forest started coming back to life. I could hear a slight breeze in the leaves. A bird chirped in the distance. Leaves crunched as a squirrel jumped off a tree and started digging for an acorn. Sometimes I wonder if Mother Nature just needs a reset of sorts. Maybe this was one of those times.

Years later, we would have a similar experience. Nate was with me and a landowner on their property as we were doing our walk-through and inventory. It was overcast, but the forecast was not calling for any rain or storms. The three of us noticed that it was getting darker and quieter. Maybe some rain or storms were forming and coming in. There was no cell service, so we couldn't check any of our weather apps, so we just kept working. The forest was very quiet, other than our talking. Eventually, things seemed to brighten up, and we took notice of birds chirping.

It was only later that evening, while I was on the computer, that I realized there was a partial solar eclipse where we were working. Thinking back, it all made sense. Animals can sense this phenomenon and quiet down. Mother Nature just seems to take a quiet pause. It was rather interesting to have worked through that without realizing at the moment what was happening, and, no, it does not explain the first occurrence, as that was a sunny day and there was definitely not any eclipse going on.

As I related in my previous book, there was a time I went looking for something I probably should not have been looking for while in the Pedro Mountains, while on a fire assignment. Again, things got eerily still and quiet for a few hours. I really do wonder if "something" was at work. Was there someone guiding my steps to keep me away from an unwanted discovery? Or, maybe it led me to the dead, almost mummified-looking mountain lion as a warning? Regardless, the world did not get loud until I was heading back. That was an eerie day for sure.

Sometimes the eerie days are kind of humorous. I am always looking for the weird or unusual in nature. Years ago, somewhere in the middle of nowhere, St. Clair County, I ran across a tree that I call my "moaning ghost tree" because, well, it looks like it is a moaning ghost. It was a day with heavy overcast and the threat of storms, yet it was still and quiet when I took the photo of the tree. I would run across that tree over the years, but it eventually failed dramatically, shattering into many pieces during a storm. I kinda miss that tree.

The last two stories I will tell come back to wind. First, there was the major storm system I was working on. I should say working between as the wind was so dry and fierce that Kansas and Oklahoma were experiencing record-breaking wildfire outbreaks. Even though it was clear where I was in the Southwest part of Missouri, the smoke and dust in the air blocked out the sun. Just letting the truck sit for a while, and there would be a layer of ash and dust. Just to the East of where I was working must have been the dry line, as that was sparking up major supercell thunderstorms and tornadoes. It was really windy, and I probably should not have been in the forest, but I was just glad not to have to deal with cither the wildfires or the tornadoes.

This final one, Nate was with me as we were working on some forest inventory. Due to some scheduling setbacks, we were in danger of getting behind on this particular project, and even with predicted high winds, we decided to get to work anyway. It always seemed the winds were not as bad as predicted, and also, being in the woods, the winds were buffered as well.

The first hour or two, we felt we dodged the bullet. The wind indeed was lower than called for, and it seemed the trees were buffering it as expected. However, the wind started to play games with us and got real gusty. The gusts seemed to be very narrow. We were on the top of a somewhat narrow ridge, and the wind gusts were blowing like crazy on one side while the other side didn't seem to have any wind at all. This blew for a minute, then it was like the wind switched sides, and what was earlier still now had the extreme gusts while the other side subsided.

There were a number of times that we stopped what we were doing and just watched the trees, making sure we would see if any trees decided to break or fall, and holding the hats to our heads so they didn't blow off. You could hear the wind hitting other ridges and then moving over to ours. Again, we stopped. I think the saving grace is that the trees had not leafed out yet, but they were sure bending over regardless. Of course, this happened as we were the farthest from the truck as possible, so the thought of stopping and heading back was hard to deal with.

Normally, I think we would have just called it a day and figured we would come back on a less windy day. But there we were. Up against a self-imposed deadline, not wanting to walk the mile or so back without getting some plots done. We stayed and kept working. A handful of times, we had to stop and keep watch for possible trees breaking or falling near us. We did hear a few come down, nothing real close, but close enough to hear, even in the noisy environment we were in. It was then that we talked and decided this was one of those "Don't tell Mom" moments. In fact, when she asked later how the day went, I think Nate said something like, "Oh, you know, just another day."

My old friend, the moaning ghost tree.

Chapter 13

Forestry and FFA

I have been extremely fortunate to work with the FFA and their Forestry program at all levels throughout my career. While I knew about the FFA Forestry team in high school, I didn't get much support, probably due to no one else having any interest in trees or forestry. Being a town kid, I really had no other interest in FFA other than being able to do the forestry thing, so my high school FFA career was short-lived.

As soon as I got to the Kansas City District, I learned that the National FFA Convention was hosted right there in Kansas City, and bonus, Larry and the Kansas City District Forestry staff were responsible for setting up and running the National FFA Forestry Contest! As I was now a member of the District, I was right in the middle of helping set up each portion of the contest.

For the thinning part of the contest, we went to an area full of younger, pole-sized trees. We had to designate a number of trees with flagging and numbered each one. Then we developed the "thinning prescription" that the students would read and use to determine which trees to cut or deaden and which ones were to remain. The students would need to put on their score sheet either "cut/deaden" or "leave" for every single tree we had flagged and numbered.

Moving on to the Timber Cruising portion, we again flagged and numbered a group of trees. This time, they were bigger trees where each tree would have to have its species determined, diameter, and merchantable height measured, and board foot volume calculated. Our job was to do this as well, to set the score sheet answers that each student would be graded against.

The Tree Identification portion was pretty straightforward. We took the official species list available to us from the FFA and went and found, flagged, and numbered trees for each student to identify. The score sheet was a listing of all the species, and the students just had to put the tree number next to the species on the sheet.

While there was really nothing for me to do on the written test, as it was set up by the larger FFA Forestry Committee, I did help with the Tool Identification portion by setting up the room and placing a numbered card next to each item. This had items such as a diameter tape, cruising vest, prism, drip torch, and clinometer. Again, I used the list to set the answer sheet by placing the correct number next to each listed item.

Of course, there were other portions, but with a large cadre, jobs and tasks got divided up. During the actual contest time, we would be out in the areas as monitors to be sure there was no sharing of information or cheating. There were a number of questions asked of us by the students, but we couldn't really talk to them.

Seeing the teams from each state come to the contest was rather inspiring. Most of these kids were really serious about this contest, and the knowledge they possessed was impressive. Seeing this level of commitment from these students created a little resentment that I was not offered the chance to compete at this level while I was in FFA and high school. Oh well, being able to help these kids along at this point would grow to become a very rewarding part of my career.

At this point, I will point out a clarification. I mention that the vast majority of students who showed up to the National Contest were serious. There were, however, a few individuals who I wondered how they got here. They just didn't take things seriously, didn't have the right pants or footwear to be in the middle of the forests of Missouri with the ticks and chiggers and sticktights. Thankfully, this was a very small percentage of the students.

I was able to help with the National Contest for 3 years before it was moved to Knoxville, Kentucky, and I consider this one of many career highlights. After the National Contest moved, I was able to stay involved by helping with a couple of Missouri State FFA Forestry Contests and a lot of local District Contests. While working out of Clinton, I set up the District Forestry contest each year at the Amarugia Highlands Conservation Area or at Settles Ford Conservation Area, both near Archie.

Everything was basically the same as I had set up during the national contest. Having the knowledge and experience of the national contest made this an easy task for me. I think that the only difference was that I had to use more of my own tools for the Tool Identification portion. I didn't have a complete set of stuff provided to use and ended up using my own cruising vest, prism, clinometer, pulaski, drip torch, and others. Yeah, you get the idea. I would go set it up one day and then be back the next for the District Contest and take down and grab all my stuff.

Over the years, I also had the opportunity to work with a number of high schools and their Forestry Teams. Some would invite me for a meeting to discuss how the contest worked or what to be sure to study for. Some had me come to work on tree identification or cruising, or how to use the tools correctly. The vast majority of assistance was a single meeting to a few visits at most.

There was one group, however, that really poured a lot of effort and commitment into learning. It started off innocently enough with a quick meeting one evening, where I was asked to talk about the ins and outs of the contest. Like so many before, I figured this was a one-and-done. I was pleasantly surprised when they called back and asked if I would be willing to coach them up and really invest in their learning. This developed into twice a week, after school training sessions. We were able to get deep into tree identification, really get into the how and why of forest management, and how that impacted thinning decisions. We cruised and measured hundreds of trees to refine their skills.

All in all, this was a great group of students who showed me a lot of commitment and integrity. When one student wasn't able to make one of the training sessions, the others made sure he was caught up on what he missed. They worked well together, were respectful to me, each other, and everyone they came across. To call them kids is really not fair; they were the kind of young adults we hope all kids turn into.

I had a great time with them and their Advisor. I didn't really think much about it, but when I got an invite to attend the Annual FFA Banquet, I thought it was a nice gesture. I didn't really think of going, but when not only the advisor, but each of the Forestry Team members contacted me and asked me to attend, I figured I had better go. Besides, it was guaranteed to be a good catered meal.

You can also imagine my surprise when, later, after the meal, I was asked to come up front. They presented me with an Honorary Chapter Degree. While some may think this is small beans, just an Honorary Chapter Degree, for me, it was special. Special because of the group of people who were behind it. A shared bond and mutual respect. I was just doing what I do, helping teach interested individuals about something I do every day. They saw it as above and beyond. I truly do appreciate that group of young adults and know they are all on their way to an awesome life ahead. For all the years I put into the FFA Forestry Program at all levels, the small acknowledgement that came from the Lakeland FFA, and the Lakeland FFA Forestry Team meant the world to me.

These kids from the Lakeland FFA Forestry Team were a great group to work with. I appreciate each and every one of them. Photo by Tammy Shroyer.

Chapter 14

From Planting to EAB, Nice Ash

Ash trees are actually one of my favorite trees. Not a top-five tree by any means, but still a favorite tree nonetheless. There are a lot of things to like about Ash. They grow tall and straight, have a pretty fast growth rate, and they tend to survive pretty well as newly planted seedlings.

Common names for ash are rather interesting. It makes me wonder who was responsible for naming them, as they seem to be either colors or places. You've got green ash, white ash, blue ash, and black ash for the colors. Then there is Arizona ash, California ash, Carolina ash, Chinese ash, Chihuahua ash, European ash, Himalayan ash, Japanese ash, Mexican ash, Oregon ash, Texas ash, and Tropical ash for ash named by places. While I don't know the history behind all these ash, it just makes me wonder, were we just lazy when it came to naming the ash trees? Hmm, a new ash tree in Oregon, guess we'll call it Oregon ash. Hey, an ash in Texas, yep, Texas ash. Regardless, most of my experience lies with the colored ash trees, specifically green, blue, and white ash.

Early in my career, when the Conservation Reserve Program (CRP) was in high gear, many farms and fields got planted to trees, and I was a part of that. The program requirements stated that at least 200 trees per acre must still be alive after three growing seasons. Generally, most of these projects were planted at a rate of 302-435 trees per acre, depending on a few factors. We always tried to get a good variety of species and took into consideration what the landowner wanted to do someday. Many wanted walnut as that was a high-value species, both for the nut meat and the lumber and possible veneer products. Others wanted lots of oaks for acorn production to attract deer and turkey.

We learned early on that some species of seedlings survived better than others, especially when planted for the CRP program. Yes, there were site prep requirements and standards that said the newly planted trees must be cared for over the first 2 years, but the reality was that these fields were being taken out of production, and therefore, they really weren't thought about much after that. In addition, it seems like most of the fields going into trees were highly erodible, low-elevation, wet fields. The oaks were great, but grew slowly, and the deer seemed to like to browse them first. Due to known mortality rates for planting small seedlings and then letting them fend for themselves, high planting numbers are required.

Still, some years we had untimely wet or dry weather, and mortality seemed to be too high, and we struggled to maintain the required 200 trees per acre. Luckily, ash, and more specifically green ash, seemed to survive well regardless of wet or dry weather. Enough ash would survive along with the handful of other species that we were able to consistently meet the survivability standard at the end of the 3 years. In fact, looking at some CRP plantings 10, 15, and even 20 years later, ash was a huge percentage, if not all of the remaining trees. Maybe there would be a random surviving oak, walnut, or pecan from the original planting. Sometimes, natural ingrowth would allow neighboring oaks or hickories to gain a foothold as well. But, in my opinion, ash is what made that program as successful as it was.

This ability to survive planting in multiple environments also made ash a good tree for urban plantings as well. With the bright yellow fall color of green ash and the deep purple of white ash, numerous varieties of each were being planted in communities all over the country. Many were being overplanted. This, unfortunately, would lead to a dark chapter.

In 2002, near Detroit, the Emerald Ash Borer, shortened to EAB, was first identified. If you haven't heard of it, this is a highly invasive pest, native to Northeastern Asia. While it was first identified in 2002, some suspect that it may have been here as early as the late 1980s. Our native ash trees have no natural defenses to this invasive pest, and tens of millions of trees have been killed. The pest outbreak spread rather rapidly out of Michigan and was officially identified in Missouri in July 2008.

Similar to Dutch Elm Disease that took out millions of large, beautiful elm trees across the country from the 1930s through the 1970s, EAB is wiping out large swathes of ash. Over my career as a forester, working with native ash out in the forest as well as urban planted trees, I have seen the total and rather quick devastation firsthand. Looking back at some of the CRP tree plantings from the beginning of my career, they are completely dead and falling over. Doing forest inventory, miles from any community, entire stands of ash have been wiped out. It is a rather stark reminder of the power of invasive pests.

Even more damaging is when communities I have worked with over my career have dealt with this pest. When a community has over 30, 40, or even more percent of its trees as ash, the loss can be almost unbearable. Imagine your community's prize city park, made up of almost all ash, and not just any ash, but large ash. I'm talking about trees 70 to 80 feet tall and 4 feet in diameter. Imagine the shade those trees can provide. Now imagine them all dead and needing to be removed over the span of 2-3 years.

Try as we might, most communities can't pivot as fast as this pest has moved. The best defense, of course, is to have a high diversity of species so that when something like this happens, you are only losing a small percentage of your community's trees. As of this writing, EAB has jumped into Texas, Colorado, and Oregon. I work with a number of communities that have not yet had to deal with EAB, and I can find myself getting frustrated when we talk about the high percentages of ash and starting to plant other species, and yet nothing seems to be happening. The wave is coming; it is just a matter of time. So many communities are going to lose their ash trees, and that will be anywhere from 20 percent to almost half of their trees. Are you ready to lose this much shade in your parks and back yards? Hmm, you'd better be planting trees now to help offset this inevitable outcome. Gone are the days of hearing "Nice ash."

Looking at an ash that has had EAB in it for
a while. Photo by Tammy Shroyer.

A stand of mostly green ash that is dying due to EAB infestation.
It is sad to see this westward expansion of this pest.

Chapter 15

The Long and Short of Flagging

I don't use flagging for every job, but I do use it quite frequently. I mean, any forester who spends time in the forest has to from time to time. I can't tell you how many rolls of flagging I have used over my career, but I know I can measure the total in dozens of miles. In fact, one recent order for a specific project was 12 cases of flagging that totaled over 8 miles. That was 144 rolls of flagging, and I only had three rolls remaining at the end of the project.

Early in my career, working for Larry on the Kansas City District, I saw one extreme of flagging usage. In this, I mean, we still had to use flagging to mark out potential trails or harvest roads; however, Larry's idea was to conserve as much as possible. Remember, he was super tight with his budget, and therefore, the flagging budget was tight as well. Larry's philosophy on flagging was to use no more than eight or so inches and tie it to a small twig. Sometimes it looked like a bowtie on the twig; it was so little flagging. Of course, we went round and round on usage.

My philosophy is to use more, like a long tail that flutters in the wind, then you can spread out farther on hanging your flagging as it is far more viable for much farther distances. In the end, it was probably about the same total amount of flagging used, whether it was a bunch of short flagging tied close together or fewer but longer pieces of flagging.

Of course, doing my own thing now, I have full control over how much flagging I use, and I do prefer the longer tails drifting in the wind. However, when I get down to the last bit on the roll and I end up with a very short strip of flagging, I fondly think of Larry and tie it on with a smile.

Flagging has changed over my career as well. When I started out, flagging only seemed to come in a few standard colors like orange, blue, and yellow. You really had to look to find other colors, and most of the time, they were more expensive and harder to get in bulk. Along the way, specialty flagging came along, specific to our industries. In forestry specifically, we started getting flagging with words printed on it, such as "Skid Trail", "Harvest Boundary", "Do Not Cut," "Riparian Management Zone," and so forth. Additionally, within the wildfire part of the job, we have "Spot Fire", "Escape Route", and "Killer Tree" flagging.

Most of the flagging we use is vinyl and is meant to last for years, but nowadays, you can also buy flagging that photo-degrades in about 6 months. This is great, until the project gets delayed for some reason or another. They even have biodegradable flagging, but that really doesn't last very long and is pretty pricey. Next, for working during the coldest days of winter, cold-hardy flagging resists getting brittle down to -40 degrees Fahrenheit. And, yes, you can tell the difference when it gets that cold.

We also get so many more colors of flagging these days. It is nice to have a number of fluorescent colors to choose from, but I really haven't found a need for flagging in solid black. Striped and checkered flagging has come in handy a time or two, but green flagging during the summer and brown flagging in the fall seems a bit counterintuitive. Then, of course, there is white flagging in the winter!

While it may seem pretty boring, all this stuff on flagging, there are times when you can get a laugh out of it. One time, I was in a neighboring district helping on a project, and there were instructions to just follow the flagging to get to the job site. Well, that was all well and good until there were four different colors of flagging leading four directions from the parking lot. That made for an interesting morning.

Years later, after I moved to Wyoming, flagging would continue to provide some interesting stories. First, this is when I really understood the benefit of cold-hardy flagging. Another day, I was going out to find the flag line that my predecessor had hung to delineate some boundaries for a larger fuels mitigation thinning project in a stand of mixed Douglas-fir and lodgepole pine. Reading through the notes and looking at the map, this seemed a pretty straightforward task. After all, I just wanted to see it for myself before we moved forward on the project. However, I was not able to find any flagging. I bounced up and down the slope to make sure I didn't just miss it. Nothing. This was odd as flagging usually lasts for at least a couple of years. Defeated, I headed back.

After talking with Brian and Ron, my Assistance District Forester and the local Hazard Fuels Coordinator with the Fire District, we all agreed that it must be there, and maybe I just wasn't finding it. Of course, that started the teasing of me being new to Wyoming, was I really a forester, did I really know what flagging was, and so on and so forth. So, we all headed back to the site a couple of days later. After we all three walked the same area for about an hour with no flags found, I was feeling a bit better that I hadn't just somehow missed them. Discussion turned to Paul, the previous District Forester, whose job I now had. Paul was extremely reliable. If he said he did it, he did it.

Now, I don't know what actually made me look up. Maybe it was a raven or a jay that called out. But, as we stood there wondering what the heck, I looked up, and there, about 25 feet off the ground, was a piece of blue flagging hanging from a branch. Wait, what? How in the world did that get there? We all looked and were figuratively scratching our heads. Brian, who stands about 6'4 or 6'5, finally said what we were all thinking. How could Paul, who maybe stands 5'5 in his boots, reach that high up? I think at that point, Ron asks a question that gives away his thinking. When did Paul put this flag line in? A quick check of the notes revealed it was late January. Now, things are becoming clear.

In this part of the Wind River Mountains, on this aspect, the snow can get fairly deep some winters. As it was, we all finally figured it out. Paul, light as a feather and with the aid of his snow shoes, put the flag line in while walking on 20 feet of snow! We started following the map, and sure enough, there was a row of blue flagging, spaced appropriately, but all about 20 to 30 feet off the ground. I was in the right area the whole time. I just didn't look up. Well, a good laugh was had by all, mostly at my expense and the occasional comment about how Paul had grown some long legs right before he retired.

In the end, we re-flagged the line as the project would be getting started before the snow of winter came along, and it might be a bit of a stretch to ask the contractor to follow a flag line that was 25 feet high. All in all, flagging, which is a pretty boring yet important tool in our toolbox, has provided some interesting stories over my career. Now, when you are walking in the forest and see a strip of flagging hanging from a branch, I bet you will ponder a lot more than you ever dreamed about, about the how and why of that little piece of vinyl.

This is just what Larry would want in a flag line. He would be so proud.

Tammy is getting the hang of using longer flagging.

Flagging, flagging and even more flagging. How many do I need?

Chapter 16

Don't Touch My Snacks

One of the first things anyone who works or plays outside learns is to always have a good supply of snacks. You know what I am talking about. The granola bar or trail mix in your pack is almost as important as the water you carry, right? Well, this has been no different for me through my career. Even now, regardless of work or play, if you look in my vehicle, you will probably find a couple of sticks of jerky or a half-eaten bag of chips. Earlier in my career, however, I carried an assortment of food and snacks that could stock a small grocery store.

Although this starts with the wildfire part of the job (you can read my first book, Show Me a Firefighter), I include this here as food is an everyday necessity. But, back to the wildfires. Many days, we would get called before lunch or work past supper, and even though we might have packed a lunch, some days just went longer than expected. The amount of food we kept in our vehicles started to increase to account for those longer days. We also needed food that would not spoil, so there could be all manner of canned food.

There were some occasions where someone would show up with a number of pizzas or a box of cheeseburgers donated by one of the local restaurants, but these were few and far between, and sometimes we were so far from town, no one knew where we were anyway. In addition, we hosted many folks who came to help out during our bad fire days, so we all carried extra food for them as well. Everyone had their favorites. Some preferred fruit like apples, oranges, or bananas, but those were short-term and were in big danger of getting squished. Others liked granola bars. Some liked a stash of Little Debbie oatmeal cream pies. My go-to was canned peaches or pears, diced into small squares, pop-top of course, and those oatmeal cream pies. Fruit and sugar were great for a quick boost of energy, but we also really needed protein on those long days.

Canned tuna, beanie weenies, and similar items were the most sought-after snacks when a break in the action allowed, and while we all shared as needed, it seemed that my stash kept getting hit harder than others, especially in the protein aisle. So, I started trying to figure out what I liked that no one else would touch. The answer? Sardines. I have loved sardines since I was a kid, and really thought the key and roll-up tab to open the can was kinda fun as well. Bonus for me? No one else on the crew liked or could even stand them, really. These never got stolen from my truck, and I would always have a supply of protein from then on. I would gladly share anything else. Cans or pouches of tuna? Sure. Some of my fruit cups? You bet. Crackers or an oatmeal cream pie? No problem. But don't touch my sardines.

To mess with some of my crew, I always offered the sardines first. That resulted in a rash of threats and insults, and fake gagging. Only after this would I offer something else, but I was usually met with an untrusting glance. So, it was early on that we all learned each other's preferred tastes in snacks and what was considered "community" snacks versus an "untouchable" snack. This should not be confused with Uncrustables, although those are "untouchable" in my book, especially now that they came out with raspberry.

These days, working mostly with Tammy and Nate out in the forest, most everything is now considered "community" snacks. Thankfully, we all like the same things. Also, since we are now in more control of our schedule, the extra-long days or overnight adventures are a thing of the past. However, if you were to run across any one of us in the woods, dollars to doughnuts, we will have at least one snack stick from Nadler's Meats, and probably a handful hidden away somewhere in our cruiser vests. Growing up in Wellington, I have known Nadlers my entire life. If you haven't had some of their award-winning products, you really should give them a try. They are definitely a go-to snack for me even today. Just get your own. Don't touch my snacks.

Tammy has her own version of "don't touch my snacks" in
her cinnamon popcorn from Topsy's.

Chapter 17

The Redwood Healing

There was this particular fire I was on in California, but this story is not about the fire. It is about what happened after the fire. I had flown into Sacramento and had a rental vehicle, and since I was done and released from the fire, and had a full extra day before my flight home, I figured I would do some exploring.

Since I was in Northern California, having been around Happy Camp, which is not always happy if you are there on a fire detail, but then I got moved closer to Idlewild. This actually made my excursion a bit easier. Once I was officially released from the fire and in travel status, I took off and headed to Crescent City to catch Hwy 101 and head down the coast. I had two goals: stick my tired feet in the cool ocean water and spend some time seeing the redwoods.

As far as the first goal was concerned, once I turned onto Hwy 1 and made it back to the coast, I found a beach where I could park. I was still in my dirty fire clothes and fire boots with three-day-old socks on. Just the thought of that cold water on my tired, achy feet was almost more than I could stand. I knew I had a towel packed somewhere in my gear, but I was having trouble finding it all of a sudden. The more frantic I looked, the more frustrated I became. I didn't want to wait any longer, so I grabbed a dirty t-shirt off the top and took off, out across the sand.

I would say I ran, but that was far from the truth. My back was in serious pain as I was fighting a flared-up slipped disc in my lower back and could barely stand up straight, but I will get to that in a bit. So anyway, after I shuffled as best I could to where the sand started getting wet, I stopped and took off my boots and socks, rolled up my fire pants, and headed out into the water. Oh, that cold water felt so good. The beach was extremely flat, so I was able to get out a ways and still only be knee deep.

Whatever swelling or achiness I had in my feet and lower legs was washed away with each gentle wave of cold water. I just stood there soaking it all up. Literally, Pun intended. You hear of pro athletes who get to take ice baths after a hard game or intense practice, but I don't see that as an option for firefighters. This was a welcome alternative, however. I continued to stand there and just take it all in. There were only a couple of other people out there that morning. A younger couple walking hand in hand, an older gentleman walking his older dog on a long leash. I could have stayed there all day, but a nagging flight schedule meant I had to get moving. Back to my boots, and I used that dirty t-shirt to wipe the sand off my feet before putting those wonderful socks and boots back on.

As far as the redwoods go, I had seen some the summer I turned 16 while visiting my aunt and uncle who were living in Tracy at the time. I really couldn't remember much about those trees, you know, typical teenager indifference or something like that. Anyway, as I made my way down the Northern California coast, I stopped at a few places. No, I don't remember the names of the parks or groves I visited. I was looking to avoid crowds as I wanted to just spend time with the trees. You see, fire clothes are pretty identifiable, and "the public" WILL ask endless questions.

At one of the places I stopped, I was able to walk around and see big, beautiful redwoods. It was so shady there on the forest floor. I was able to just enjoy feeling small and insignificant next to those giants. Funny, as much as my back hurt, I was able to move around a bit better. I had a small tripod with me, so I was able to set up my phone on the timer and take a few staged photos. I'm sure that to anyone who was watching, it must have been quite the spectacle. A hobbled-up guy setting the timer on the camera, then trying to get in position as quickly as possible, and trying to stand up as straight as possible before the shutter snapped. I think I did get a couple that worked out.

For me, this was the quiet downtime, the healing I needed after a long, hard detail fighting fire. No radio squawking, no generators running, no helicopters flying overhead. Just quiet. The sore back and aches from sleeping on the ground for a few weeks started to ease up and maybe even melt away.

The mental quiet was also welcome. At the moment, I only had myself to worry about. No longer was I in charge of other firefighters who were carrying saws and tasked with cutting down the dangerous, high-risk trees and snags. It is not that I was only thinking of myself; I wasn't. I was looking forward to getting back home to the family. However, not every fire or every trip allows for some "self-help" before you get home. I was going to take the full opportunity to "heal" myself as much as possible before I got home so I could give the family my full attention.

Spending these extra hours in the forest, alone and among giants, I was able to just feel the vibrations of the earth and the trees. Like I said, it seemed to reduce the pain and soreness I was feeling. I was able to process and compartmentalize recent events so that there wouldn't be excess baggage at home. Now, somewhat refreshed and rejuvenated, I resumed my journey to Sacramento, turned in my rental, and caught a shuttle to the hotel. Only after a very long shower on a continuous lather-rinse-repeat cycle did I pull out the hidden pack of clean clothes I saved for the trip home.

The flight home the next morning was uneventful, and it was so good to see the family at the airport to pick me up. As no one else in the family had seen redwoods yet, I started in on the stories. They kept asking about the fire, but I kept coming back to the trees. I guess that is what had the biggest impact on me during that particular trip. For someone who spends so much time around trees and in the forest, I think even I can forget to slow down and just be in the forest. Just be. Let the trees, let the forest, heal you from whatever ails you. Just let that sink in.

One of my timer photos in the redwoods. This was so refreshing to be amongst giants.

Oh that cold water felt so good.

Chapter 18

Hey Bear,
(and Wolves and Cats, oh my!)

Most foresters will work alone at times. In fact, it seems that a large portion of our career is spent working alone. I rather enjoy the solitude and quiet that comes with working by myself and have never really thought about it. Yes, there are safety issues and concerns as you have already read about, but animals didn't cross my mind. That is, until I moved to Wyoming.

Let me back up a bit. Spending time in the forests growing up near the Missouri River, I never had a concrete thought about not being at the top of the food chain. I always had my trusty knife, and many times I carried my .22 rifle. Since squirrels were in season most of the year, many times I made it home in the evenings with a handful. Once, when I was about 10 or 11, there was a report of a possible mountain lion in the area. This being Missouri, mountain lions hadn't been confirmed for decades. But I was intrigued. I imagined myself like Davey Crockett, in my coon skin cap, knife in my teeth, fighting the mountain lion in hand to paw battle.

Some of the older people around town told stories of how a mountain lion sounded like a woman screaming. Well, now I was always keeping my ears open for this sound as I was out and exploring. Finally, one evening, I did hear what sounded like a woman screaming. Maybe this was the mountain lion! However, the screams were intermittent, and I could never get a precise location.

I never did run across that crafty old cat, although I did set out for a couple of days, fully intent on fighting that beast. As it was, the last confirmed mountain lion seen in Missouri was in 1927, and the next one to be confirmed would not be until 1994. I am not here to debate whether mountain lions have been in Missouri between those dates or not. All I know is that in the mid-80s, when I was on my above-mentioned quest, I was unsuccessful.

I will pause here to offer some other thoughts on mountain lions. It seems that there have always been reports of mountain lion sightings. I never really thought about it until I moved to Wyoming. There are a lot more mountain lions in Wyoming and neighboring Colorado and Montana, and yet the sightings seemed lower than what I remembered in Missouri. Maybe this had to do with differences in population, or maybe people were used to them and just didn't report them. Regardless, it is a fact of life that the local school in Riverton has had a few mountain lion lockdowns; kids could not go out for recess due to a confirmed mountain lion in that part of town.

Anyway, back to the story. I mention all this to give the context of me not really being too concerned with such beasts in the dark woods and back hollers. Even when the black bears started moving up into Missouri from Arkansas, I was not afraid of them. I just wanted to see one. Sadly, to this day, I have not seen one in Missouri, but you can read about some of my bear exploits in my previous book.

Moving to Wyoming, specifically living and working in the NW corner of the state, means a different set of circumstances. This is grizzly bear and wolf territory. I learned early on, even if you have someone with you, you need to be extra cautious while out and about. It was not uncommon for me, or if I was with someone, for one or both of us to be talking loudly or hollering out, "Hey Bear!" See, most bears would rather avoid humans, but things can go badly if they are startled. By making noise and talking loudly or yelling, we gave the bears a chance to move along and avoid the chance of meeting up. Some days, I felt silly with my continued "Hey Bear," so I would mix it up a bit. I called out Yogi, or Smokey, or something similar, just to keep it different.

Again, I really didn't worry about the black bears. It was the grizzlies that concerned me. Every year, in the area, there are multiple reports of people getting attacked, and some killed, by grizzlies. Some of these people are people I personally know, or that Tammy or Nate know, so it is a very real thing.

Wolves, on the other hand, were a fascination for me. I have heard wolves howling many times while working in remote parts of the Wind River Range or the Absaroka Mountains. I know they are there, I've seen numerous tracks, and even seen a handful over the years. One time, however, it gave me chills.

It was still winter, and I was up in the Grass Creek area on some state sections looking at some potential thinning projects in some Douglas-fir. This was when I was the District Forester with the Wyoming State Forestry Division. I had to leave the truck on the main road, and due to some fairly deep snow drifts, I had my snowshoes. I made it about a mile from the truck, and in one valley, the snow depth dropped to about 4-6 inches, so I took the snowshoes off and strapped them to my back.

Pretty soon, I came across a set of tracks. Definitely wolf due to the size, and they seemed rather fresh. "Cool," I thought to myself. Maybe I'll see it up ahead of me somewhere. Soon, how-ever, this track met up with a bunch more. Initially, it was hard to tell how many animals were in this pack, but at one point, I was confident I could separate out eight different and distinct sets of tracks.

"Cool" was not how I would describe it any longer. Yes, it was cold. It was winter after all, and there was snow on the ground. But the chill that ran through my spine was not cold-induced. Having the realization that I was vastly outnumbered, over a mile from the truck, and in an area where probably nobody knew where I was, was a bit uncomfortable. I kept arguing with myself as to the logical part of this. Logically, while these tracks were recent, they were made overnight, and this pack was miles away by now. Wolves would rather not bother a human, as humans can carry boom sticks or thunder rocks.

The primal part of my brain, however, argued that I was just a meal waiting to happen and to immediately start back for the truck. On and on the argument went. With over a two-and-a-half-hour drive to get to where I parked the truck, and over a mile hike in snow shoes and then another mile or so without, it would be a huge investment of time lost if I turned around. "So what? Who cares if you are just filling the belly of a pack of wolves?" "But it is just another few hundred yards to where I need to go, I'll be fine." "They are waiting right there in the very stand of trees you need to check out." "No, they are up in the wilderness by now."

On and on the argument in my head raged. The logical part of me finally won as I reached my destination. I checked the map and old flag line, hung a few new strips of flagging, and the mis-sion was completed. I headed back. I never did see any wolves that day. They really were probably miles and miles away, but the feeling was hard to shake.

I have a healthy respect for the natural world and the animals in it. Whether we are truly at the top of the food chain or not doesn't really matter. How we interact and cope with those situations sets our place in the order of things. I enjoy seeing all manner of critters, even the bears, wolves, and cats that could eat me if I let them. I just chose not to let them. And so, I will continue to holler out "Hey Bear!" when I need to. And maybe, just maybe, Yogi might answer me back by telling me, "I am smarter than the average bear."

Wolf tracks in the snow on my journey.

That is one big grizzly bear. Glad he is going the other
direction. Photo by Tammy Shroyer.

Chapter 19

Technology

The tools that we use in forestry are an interesting mix of antiques and cutting-edge tech. Let's take some of the common tools I carry on a typical project. First, the wedge prism. This is a small piece of glass that is thicker on one side and thinner on the other. You look through it at a tree trunk, and due to the different size, the tree trunk looks offset. We use this to determine if a tree is "in" or "out," as in, do we tally it or not on a certain type of forest inventory. The wedge prism, as we use it today, was developed in the mid-1950's so it has been around for a while.

Once we figure out which tree to collect information on, I then use a Biltmore stick. The basis of a Biltmore stick is the concept of similar triangles, so that we can estimate the tree's diameter. Since most inventory projects only need tree diameters grouped into 2-inch size classes, the Biltmore stick is great. This tool has been around since the late 1890s.

Next, when we need tree height, I grab my clinometer. The clinometer we all use today is an improved version of the Abney Level that was invented in about 1880, although inclinometers in general have been used for centuries. It is interesting that some of the very basic tools we use daily are really old devices. I mean, the math hasn't changed. Some improvements in design or function, or reduced weight, are about it.

While those tools have not changed, others have, and drastically. Even though aerial images and GIS were a part of the world when I went through forestry school, we still used a stereoscope and matched stereo photos to look at stand delineations and aspect features. Stereo photos are a set of aerial photos that are matched but taken from slightly different angles. For example, a plane usually flew a grid and took pictures of the forest at set times or distances. The plane then turned and flew back the other direction, but this time further left or right. When you paired the photos together, you could then look through a stereoscope to "see" the terrain in a three-dimensional effect. The terrain always appears exaggerated, but you could easily tell the ridges and hollers, as well as how steep the slope may have been.

These days, I have my own drone and capture my own aerial photos and even make them into new orthomosiacs. The camera technology has come so far that I can even make fairly accurate topographic maps with software that can calculate distances to a tree versus the ground from multiple photos. Other times, I use the camera on the drone to get a different perspective of a tree in someone's yard. Depending on the nature of the call, such as a sick tree call, storm damage, or a high-level tree risk assessment (ISA Level 3), I can use the drone and camera to zoom in on parts of the tree that I can't see from the ground and without the risk of a tree climber.

This has allowed me to see the tops of affected branches where I could see a canker that I could not see from the ground. I was able to see how bad a failed branch was and that it was still attached. It had not fully separated from the main tree. Another tree that I was called for was showing signs of branch dieback and yellowing leaves. Due to the height of the tree and the lower canopy being very full, we just could not see the top of the tree and the affected area very well. The drone allowed me to see the top side and determine that squirrels eating and peeling the bark during a hard winter were the culprit.

Even more, I also have a LiDAR sensor that I can carry by hand or attach to yet another drone. LiDAR stands for Light Detection and Ranging. It emits a laser light out at 600,000 pulses per second and reads the return pulse, and can therefore calculate the distance to an object with accuracy less than a centimeter.

This allows me to either capture trees in 3D if I carry it through a city park, or if I fly it over an area, I can get highly accurate ground points in order to produce a new topo map. Your typical topo map gives contour lines in 5, 10, 20-foot or more intervals. In a rather flat area, it can be hard to look at a map to determine fine details, but with the drone-mounted LiDAR, I have produced a new contour map of a city park using 4-inch contours. Talk about seeing small differences!

With the LiDAR scans of a city park full of trees, we have been able to not only have the park in 3D, but can also use that data to calculate the diameter of each tree, not only at our normal DBH measuring height of 4.5 feet, but for the entire length of the trunk. This can help visualize a number of things, like total woody volume that can be used in planning efforts to preplan debris removal in the event of a tornado or wind storm blowing all the trees over.

This can also make us more efficient as foresters by allowing us to collect even more data about each tree than we ever thought possible, and doing it in a fraction of the time. This will allow foresters to inventory more parks or urban woodlots, and the more that can be inventoried, the more that can be managed. The more that can be managed, the better off we all are.

I feel pretty fortunate to have seen as much technology change throughout my career as I have. Who knows, by the time you read this, where will we be with technology? I can only hope and wonder, but for right now, I am going to enjoy a quiet, peaceful walk through the park and slow down while under the shade of a beautiful tree.

Nates uses the wedge prism while I use the iPad. Photo by Tammy Shroyer

Setting up the drone to do a LiDAR flight.
Photo by Tammy Shroyer.

Nate using the Biltmore stick to get a diameter on this tree.

Chapter 20

Basketball Anyone?

Making your livelihood in the forests, you have some interesting things happen throughout your career. Now, sometimes that "interesting" thing is the fact that nothing interesting seems to have happened for 4 or 5 days in a row. The same species, day after day. Trees about the same size as you move from stand to stand. The same terrain up and down as you move through a compartment. You really start looking for the minute differences in each stand and start to wonder about the merits of a "lumper or splitter" philosophy.

One case in point was a rather recent inventory contract in the oak and hickory of the Ozarks. For this particular project, I found myself putting in plots with a Rinse and Repeat monotony. It seemed like every plot was 15 to 17 trees, and each tree seemed to range in size from 12 to 16 inches. About now, any forester who is reading this and is familiar with Central hardwood forests may pipe up and say, "That must be some good stands!" High basal area and all in the small sawtimber size class. What is the complaint?

Well, about 30 to 40 percent were dead or dying, and there was no apparent oak regeneration, or any regeneration for that matter. On and on this went. Plot after plot. Stand after stand. Day after day. Almost half of the compartment, or about 300 acres in this case, ended up in this condition. I really started to wonder about how this came to be.

I didn't know much about the area other than it was purchased by MDC in 1983. These trees, being much older than the 42 years since the purchase, gave some clue that whatever management set those stands up occurred much earlier. My best guess was that the whole property was cut over in the 1950s and then abandoned. It grew back in trees, as most oak forests do, and was sold to MDC about 30 years after. As the trees were actively growing, there was not much management other than some possible thinning by the state.

Next, cue in the droughts of 2011 and 2012. This started the sequence of events that led to what we had that week. Severely stressed trees were not able to fend off the ever-present Armillaria root rot. So, many years later, the stands have stagnated, and many of the trees have finally succumbed to the root rot disease.

With that little bit of background, I come back to the current state of 300 acres of no variability. Over my career, I have tended to lean more to being a "lumper" than a "splitter" in that I lump small differences in a stand, or inclusions, into the larger stand. Splitters, however, may choose to carve out those small differences. The results can be strikingly different. In this case, 300 acres from a splitter philosophy could result in 43 stands of less than 7 acres each. This could be perfectly acceptable by having a different stand for the north-facing slope, the flat ridge, the south-facing slope, and so on.

Likewise, in the lumper philosophy, 300 acres could be 10 stands averaging 30 acres. Again, this could be perfectly acceptable by having a single stand from drainage to drainage, as the north slope, the south slope, and the ridge are all accessed by the same trail running down the ridge. There are pros and cons for each way, and personal preference by every forester will dictate the final stand map. As I said, I tend to be a lumper, and I look at how the forest will be managed. If I see that 3 or 4 stands will have the exact same treatments at the same time, why split them out? Just make a single bigger stand. With this compartment, I found my mind wandering and debating what the recommendations would be. All 300 or so acres were the same. All needed to be salvaged before any more trees died. This could be one big, 300-acre stand!

Yeah, yeah. I know, rein it in. That is not practical. And let's get back on track. All this said to illustrate my point of a large area, days' worth of work, with nothing interesting to separate it out or break up the day. I mean, yes, the interesting part is the lack of something interesting.

Wait, you ask, what does all this have to do with basketballs? I'm getting there. First, let me chase another rabbit down a hole. On the other end of the spectrum, peculiar things can happen, or we may come across interesting finds. A couple of quick examples of some interesting days include a freak downpour during a day with no rain in the forecast and the resulting flash flood. Thankfully, I was on high ground when it happened. Or, the haunted, ghost face post oak I found once.

While "interesting finds" can be its own book someday, let's finally get back to the basketballs. Nate was helping me with a couple of forest inventories for MDC. We were finishing up one area and then starting a new area over the course of about 3 days. As we finished up the one area, the second-to-last stand we were working in, we came across an old basketball in the middle of the woods. Of all the things I have found in the forest, this was the first basketball for me. A quick photo to add to our collection, and we were on our way to the next plot.

In the next stand, over on the opposite side of a forest drainage, about halfway through our plots, guess what? Yep, another basketball. Ok, this is definitely interesting. We stop for a few minutes to look at the map. Was there a house just up the way where these came from? They were not down next to the actual creek, but mid-slope away from the drainage. We chalked it up to just an interesting day in the forest.

Jump ahead to the next day and a new Conservation area two counties away. In our third stand of the day, a bottomland stand, was another basketball. This one we figured did wash in on the creek during a flood and got caught as the water dropped. Nate and I joked about the odds of finding multiple basketballs over two days in a row in two different areas. This is starting to climb the chart on our Interesting Day list.

As luck would have it, guess what? Yep, we found more later in the day. And the next. Three days in a row, on two different conservation areas in two different counties, we found basketballs while doing our forest inventory. While we did not take a photo of each one, we did end up with a final tally of 14 basketballs! Never in my career had I found one, then 3 days and 14 total, and not one since. Yep, that is what makes for interesting days and cool stories for my career. And, truth be told, I'm not really that good at basketball anyway.

One of the many basketballs we found in that crazy week.

Chapter 21

One More Plot

I don't know when it happened. Can't really put my finger on it. But, one day, as Nate was helping me on a forest inventory, he asked me what part of the inventory was my favorite part? I think he was looking for an answer, like seeing how many different species of trees we encountered, or finding that one huge tree. I pondered this question for a moment before I gave my answer. I was actually torn between a couple of answers.

There is something to be said for the first plot of a project. You are at the starting line, waiting for the crack of the starting pistol. It means a new adventure. What will you find on this project? Will you be able to add a new story? Will you find something cool to take a photo of or add to your "Weird Finds in the Forest" folder? It seems there is always something new to learn, and even though at this point in my career, I have looked at literally millions of trees, the first plot means I get to see even more. The wonder of what comes next has an almost jubilant feeling.

Yes, there was pre-work before we even got to the forest, like looking over topo maps and aerial photos, estimating stand boundaries, or reviewing existing stand boundaries that have been provided. We may have a pretty good idea of what we are getting into, or have not one clue. Regardless, the time is nigh. You are in the starting blocks, ready to get working. That first plot of the project is just downright exciting!

On the other hand, the last plot of a project can make a pretty good argument for the best part. What if the inventory was in an area that just had a bunch of blowdown, and the walking was made rather difficult due to all the downed trees? That just happened on Huckleberry Ridge. What about the projects where we found the oak decline was way more advanced than anyone had realized? This happened to a client in St. Clair County. Or when it seemed that everything went wrong and slowed you down? A broken prism or a broken Biltmore stick? Bad weather like flash flooding, or smelly stagnant floods (remember Platte Falls?), or too much wind on a cold day in January, or no wind on a hot and humid day in August?

By the time you get to the last plot, you are ready to be done! You count down as you get closer to the end, and when you finally get to that last plot, a sort of euphoria creeps in. You may still be miles from the truck, but that last plot is screaming with excitement!

With all that, I gave my answer to Nate, and we discussed the merits of that answer. He then asked what my least favorite part was. Hmm, yeah, same answer, but in reverse. Let me explain.

The first plot also means the start of a long and arduous journey ahead. Maybe you didn't get the best night of sleep the night before, or you are fighting a headache. You really don't want to push through all that today, but here you are, trudging to that first plot location. I won't call it dread, but the mind can easily drift off to a different project that sounds a lot more fun at the moment, like flying the drone or presenting at a tree care workshop.

Conversely, what if the project were rather pleasant? You got to see a number of cool things, added a bunch of "Weird Finds in the Forest" photos to your collection, or saw some absolutely stellar stands of top-notch white oak. As you get closer to that last plot, you start to get a feeling of sadness. That last plot represents the ending of your time in this particular area. I have even found myself slowing down so as to make the experience last a bit longer before I have to be done and leave. Yeah, so in these instances, that last plot can be my least favorite.

With all this discussion about the bookends of an inventory project, we naturally migrated to the middle of an inventory. Day three or four of seven. There, I had not really given much thought to the favorite or least favorite. I did, however, have lots of thoughts concerning overall productivity and scheduling. Both Nate and Tammy have accused me of overthinking it. I am mentally calculating how many plots are left for the project, what our productivity has been so far, how many we need to get per day, and so on and so forth. Of course, I never really mention all this internal dialog, I just casually say towards the end of the day, "Hey, let's try to grab one more plot."

Of course, this usually comes at the end of a long or hard day when we are tired, hungry, thirsty, or whatever. I will say, this statement is never met with grumpiness, as neither of them is built that way, but I have been met with some "alternate reasoning" that they use to get the point across that the day is indeed done.

I guess I have quite the habit of saying this recently. Having our own consulting business this means that the quicker a project meets completion, the quicker we get paid and on to the next project. When I worked for MDC or WSFD, I didn't ever recall pushing for one more plot. There was always another day that could be dedicated to said project. My paycheck wasn't really affected by getting a couple more plots done on any particular day or not. That does not mean I didn't work hard, I did, but when you work for yourself, there is a perspective change.

There was a recent Contract Compartment Inventory project that I was doing. Tammy was helping for the week, and she was back at the truck, tending to some other business. I was firing on all cylinders and saw that I would be a pin flag short to finish a stand, so I called her and asked if she could bring me one more pin flag to get that last plot. As we met up on the trail, she handed me not one pin flag, but six. I looked at her and was about to say something about her not hearing me correctly, but she beat me to it. "I know it is never just one more plot!" Dang, beat at my own game.

She was correct, of course. We did have time to grab another stand that had five plots in it before the sun angle got too low. That was a rough week, and those five extra plots made the next day feel a lot easier. It was one of those "last plot is my favorite" kind of projects.

Overall, the "one more plot" has become something of a joke between us. One time, I mentioned to Nate that maybe we would be done for the day after we finished this next stand and head back to the truck, maybe grab an early supper. He just looked at me and said, kinda sarcastically, really, "Are you sure that's your final answer?" I'm sure I have downplayed this according to both Nate and Tammy. I guess they can give their perspective, their version of the story, if they ever write their own book.

So with that, I will end this story. Well, I mean, yeah. Or, maybe, just one more plot.

Long shadows mean we are getting towards the end of the day.
Maybe just one more plot? Photo by Tammy Shroyer

Sometimes, one more means we get rained on.

Chapter 22

My Own Schedule

This may hit a bit different than some of the other chapters, and that's ok. It's my book, my stories. It will begin with the idea that most folks who get into a conservation or outdoor type profession do so because they like to be outside, feel a sense of connection to nature, or a willingness to help the greater society. For a lot, it is a calling. No one gets into forestry to get rich, that is for sure. Focusing on forestry specifically, throughout my career, I have been able to categorize a few career paths. Where do you work as a Forester?

The first logical division is the public service sector or private sector. Within the public service sector, this can be further divided into federal, state, or local government jobs. Federal jobs that a forester may pursue include both forester and firefighter jobs with agencies like the US Forest Service (USFS), Bureau of Land Management (BLM), National Park Service (NPS), or Bureau of Indian Affairs (BIA), to name the most common. There are some others, but you get the point. These foresters are charged with managing federally owned forest resources.

Next, a forester can get hired by any one of the state forestry agencies. In Missouri, that would be the Missouri Department of Conservation for the most part, although there could be occasional jobs within the Missouri Department of Natural Resources. I should be able to say with confidence that every state and even most US territories have some form of "state forestry" agency, but with some rather interesting political moves recently, my confidence has dropped.

In my career, I have worked with two state forestry agencies, the Missouri Department of Conservation (MDC) and the Wyoming State Forestry Division (WSFD). I have good friends who work with others, such as the Nebraska Forest Service (NFS), the Kansas Forest Service (KFS), the Idaho Department of Lands (IDL), or the Montana Department of Natural Resources and Conservation (DNRC). Slightly different names, but all with similar missions, to manage state-owned forests and assist private landowners and communities with forest and tree management.

Other foresters in the public service sector work with some form of local government, such as a county or city government or equivalent. Again, a similar mission, manage the trees and forests of that level of government, like city or county parks, or Right-of-Way trees. While there may be other examples, you should be able to fit the grand majority of public service foresters into one of these three categories.

Within the private sector, we can divide jobs into industry and consultants. To start with the industry, these foresters can work for large or small companies whose job is to grow trees for products to feed whatever specific mill. This could be paper mills that ultimately produce the myriad of paper products we rely on every day, like the paper in the book, or the toilet paper in your bathroom. Or maybe it is to grow the 2x4's so you can build your house or that extra storage shed in the back yard. There are many companies, but I don't know most of them, so I won't call out any names.

Finally, we reach the Consulting Foresters. Even here, there can be logical groups. Some consultants work for larger consulting firms. Some consultants specialize in a small area of expertise, for example, I know a forester who only works on tree risk litigation and nothing else, while others are more generalist, or a Jack of all trades kind of forester. Currently, I am here at the Consulting Forester position, where I choose to work across many of the forestry disciplines.

All are, or can be, very rewarding jobs. I spent most of my career in public service, and most of the other foresters I know are, or were, public service sector foresters. However, for me, I believe that once I made the transition to Consulting Forester, I really found my groove, so to speak.

Here is where things might take a turn. While being a public service forester helped shape a lot of who I am today, I am done with that and will never go back. I have tasted the freedom of being my own boss and won't be chained again. While I have mostly talked about the awesome and cool things in my career, I am about to talk about "the other side" for a bit. Yes, every job has its good and bad points. Even consulting and owning my own business currently. No one is exempt.

However, when looking back over my career, I can see the stress of dealing with ungrateful public, or less than stellar bosses, caused more discomfort than the uncertainty of consulting and the ups and down of work. Yes, a steady, guaranteed paycheck, no matter the weather or amount of work requests, was nice, but as you have heard before, money isn't everything.

I am going to relay a number of examples that just made me question, is this all worth it? To start with, I got into forestry to be a forester. Pretty simple, right? Not for everyone. At one point in my public service career, I did my own 12-month job study. Twenty-eight percent of my work for that time frame was actually tied to forestry (or wildfire) related tasks. That means that seventy-two percent of my time was directed to something else. Things like catfish regulations, feral hogs, deer culling due to CWD, personnel performance reviews, and random administrative tasks. There was even a time reporting code for entering time in our time sheet due to having to double-report because of a lack of confidence in the new time reporting system!

For some employees, this was fine; they even thrived in that environment. It is just not what I signed up for, so to speak. I mean, I trusted our Fisheries Biologists to make the right decisions about catfish regulations. The only thing I know about catfish is that I like to catch them and eat them. I didn't feel I had enough expertise to talk with upset members of the public at a public meeting. Yet, that took a noticeable amount of time in my timesheet study.

Moving on to angry clients. Although it was not explicit, there was an unspoken expectation that you had to endure whatever the public dished out. Don't get me wrong, the vast majority of cooperators I worked with were great people, happy for the help and guidance. But it also only takes a few bad apples to ruin the barrel, right? I had one landowner who threatened to call the Governor, "a close personal friend," if I didn't move him to the top of my list and get my butt out to his property the next day. I had another who made it clear they were only wanting to fleece the system on a cost-share program we were assisting a federal agency with. It was very blatant and crossed the line of my moral compass, so I refused. Of course, calls were made to my boss and the State Forester, but even after the "talking to" I received, I stood my ground. I was not going to work with that individual.

There were other employees who were a nightmare to work with. Rude, condescending, and even crossing the line of harassment. There was a general lack of appreciation from a number of higher-ups, just criticisms when they didn't even have a clue what was really going on. I could go into a lot of very critical detail, but I think I may leave it for now. I'm getting a bit riled up just recalling some of the details.

I know that every job and every profession has these problems, and many are far worse than I ever experienced. I don't bring them up to complain, as I know my career has been very rewarding and pleasant. It is only to give a slight context to where I am now. Owning and running my own Consulting business, I have more control over the people I work with. I have chosen not to work with a few folks who I knew were going to be problematic. The money I could have made was just not worth the stress or stretching my morals.

I choose which projects I bid on and which to pass on. Some, while they look like they could be a lot of money, come with strings attached that I don't want to deal with, or are things that I really didn't like in my public sector days. I just deleted that request from my email and don't submit a bid, easy as that. I have even fired a client. Without all the details, it boiled down to some things that were said and done that were way out of line.

Although that was an example of a bad apple, it is rare in my consulting business. I have found that when someone has to pay for a service, versus getting it free, they tend to be more understanding and appreciative. I still lay out the expected timeframes, limitations, or constraints that I always have. When I explain that I won't be able to get to their project for three weeks and explain my fees for the project, they seem ok with that. No problems. Back when the service was free (as a public service forester), that three weeks would be challenged. Couldn't I move them up? No, there are other projects that are in line in front of you. People who called before you.

Overall, the stress level these days is much less as I am not forced to work with those who cause problems, are always unhappy, or are just looking to cheat the system. I can weed them out, or even if I get halfway in, I can always send them somewhere else. I am the boss; I can make that decision.

Another thing about having my own schedule is the freedom to work extra-long hours or sleep in, whenever I choose. Many times, when I travel for a project, I end up working long hours. A recent project had me driving 16 hours to get to the project, and then I proceeded to work about ten or eleven hours every day until the project was done. Seven days on the project. Was I tired? Yes. Did I have a boss complaining to me about overtime hours or comp time earned? No. Did I hear complaints about not prioritizing a different project over this one? Nope. Did I hear complaints from the client about getting the job done ahead of schedule? Nada. Did I hear complaints from my family for getting home a few days ahead of schedule? Yeah, right! They were glad to have me home early. Did I worry about sleeping in the next day and taking a "day off"? Not a chance.

About now, I am reminded of a saying I hear in business circles. A business owner will work 80 to 90 hours a week to avoid working 40 hours for someone else. Yep. There are those weeks, like I just related. But there are also the weeks when I choose to work less so I can be available for my family for whatever reason. And, as luck would have it, with this being a family business, we now collectively make those decisions to work either long hours or only a couple of hours so we can go play or rest. Some call that team building. It is, in a sense, but the family bonding, or investing in your family, sounds better to me. And, so, I will continue to create my own schedule.

A good example of this occurred when Nate was helping me on a forest inventory project. Rather, we had just finished up one project and were going to start a new one. We finished up before noon and had an hour or so drive to the next location. After a bit of discussion, we ended up going into town and had a two-and-a-half-hour lunch at a little hometown Mexican restaurant. After grabbing some dessert and some caffeine from a gas station on the way out of town, we went to get the new inventory started.

Another great example was the time Tammy and I spent about 23 days on a business trip. This was the result of projects in 3 different states, but the cool part was being able to take a few extra days and visit some of her family and friends. We had a tree damage appraisal in central North Dakota. This is also where Tammy grew up, and we made our schedule with a number of extra days so we could visit some of her extended family and cousins in two different parts of the state. We also stopped to visit the parents of her childhood best friend in Mandan, and then we were able to spend a few days with her childhood best friend on their ranch north of Dickinson.

Having the ability to create our own schedule has allowed for some pretty cool moments. Whether it is a long lunch break or extra days between projects, the joy of working together as a family, or dedicating time in the middle of the day, is worth it. And with less stress.

Having my own schedule has allowed me to see Tammy back
on a horse in the prairies of North Dakota.

Creating our schedule as part of a family business is the best.

Chapter 23

Scout Camp

Anyone who is from the Western part of Missouri or the Eastern part of Kansas will probably know about the Bartle Scout Camp. But, before I get to camp, I think there needs to be a short history lesson on the man for whom the camp is named.

H. Roe Bartle could be argued as a larger-than-life kind of person. In short, he was a 2-term mayor for Kansas City, a beloved Scout Executive, businessman, philanthropist, and public speaker. Bartle was affectionately called "The Chief" because of his leadership with the Boy Scouts and where he created the Mic-O-Say honor scouting camping program. This came about from his time in Wyoming as a Scout Executive and his friendship with the Northern Arapaho Tribe. He was called Chief Lone Bear.

When, in 1962, he helped entice the then-Dallas Texans football team to come to Kansas City, owner Lamar Hunt renamed the team the Kansas City Chiefs after Bartle's nickname. How many of you knew that history? There is so much more to H. Roe Bartle, and I encourage anyone to dive deep into his history and what he did for so many people. Regardless, his legacy lives on through things like the Bartle Hall Convention Center, the H. Roe Bartle Scout Reservation, and the Kansas City Chiefs.

Back to the camp, Bartle Camp has been a staple for most of my life. From my first day as a new Boy Scout attending the 10-day camping session until today, I have had direct interaction with that place. At the tender age of 11, getting to camp in the great oak forests near Osceola, I was hooked. Remember, at age 12, I knew I wanted to be a forester, so this was another awesome opportunity to continue to shape my future.

After I graduated high school, I was still able to stay a bit connected as my younger brothers went off to camp each summer, and we would go as a family for Family Sunday or some of the special Council nights. Of course, when I moved to the Clinton location as a Resource Forester in early 1999, my involvement took on a whole different look.

My predecessor, Mike, now my boss and the Regional Forester, had worked with them to get a timber sale going. The trees were already harvested by the time I got there, but I was able to close out the sale with the final inspections. Thus began a long journey as a professional forester with the Camp I loved so much.

Throughout the years, I worked with a number of Camp Rangers and Scout Executives, assisting with the overall forest management of the place. Since the property was larger than we were typically allowed to work with in MDC, I worked with the Scouts to contract out the development of a Forest Management Plan and did the contract compliance and quality control of that first plan. With that plan in place, I then assisted them in becoming Tree Farm certified.

Other projects I was involved with included demonstration burns as we tried to reintroduce prescribed fire to some of the glades. We marked cedar for removal and showed how to put in the correct fire lines and how to conduct a burn safely. I helped the summer staff by teaching them all they needed to know so they could be merit badge counselors for the Forestry Merit Badge.

There were a number of thinning contracts I helped them with by writing up the contract, marking the stand boundaries, marking the trees to be removed, and signing off once the work was completed. One thing that was glaring in its absence was stands of new trees. All over the property, everything was effectively the same age. It took a bit of convincing to the Council leadership at the time, but I was successful in getting them to allow three areas where we were able to slash down everything and truly get the next generation of oak trees started. This was a first for the property and a great example of working together to start the property on a sustainable future.

Years later, it was time to update the forest management plan, so I again assisted them with contracting out the project. Due to some changes in leadership, we had to make some concessions, and the direction for the plan changed slightly, but it was completed nonetheless. Already, I had hiked about every square foot of the property multiple times, whether as a kid or as a forester. This really was a special place.

As with any place that has a lot of foot traffic, some issues were becoming apparent and reaching critical levels. Between scouts, leaders, and family, there are over 10,000 people who come across Bartle each year. This much foot traffic can start to have an impact, and due to soil compaction in some areas, trees were suffering and dying. Some of the trails were having erosion problems, and there were no new trees growing in the campsites. Through many discussions with many folks over many years, the Scouts made some incredible and positive changes. Trails were better defined to keep foot traffic confined. Better steps and water bars were installed. Erosion control measures were incorporated. In the campsite specifically, "protected" areas were defined so that not only could new trees grow, but soil structure could also start to recover.

Hazard trees were identified and removed, and an overall education about the longevity of the camp was started. I will not claim much credit, other than to say I was consulted and offered lots of thoughts and opinions on the matters just discussed. It was nice to see so many positive things come from open dialog and see how they are not only preaching, but actively practicing a "living in harmony" philosophy all over the camp.

I was a bit concerned about losing touch with the Camp when we moved to Wyoming. However, we did send the boys back to attend camp each summer, so I was able to stay connected. Once I started my consulting business, I again started talking with Scout Leadership about the merits of active forest management. It took a bit, but we finally reached an agreement to get back to active management.

This is the third iteration of a forest management plan that I have been involved with at the Scout Camp, although this time, we are taking a more holistic approach. We have broken the entire property up into five compartments and are taking one each year. This involves new stand typing, conducting the forest inventory, and developing the forest management plan. While it may seem to be a little slower process, it is a more in-depth and detailed process.

Currently, we have recommended timber sales, not just to start a little cash flow, but more importantly, to regenerate the oak history forests that so many people have come to know and love. There are thinning projects. Glade and Savanna restoration with prescribed fire. Finally, there is an all-out assault on maple and cedar.

The biggest threat to the sustainability of the forests that are and surround the camp is the growing abundance of sugar maple and eastern redcedar. These are species that can handle more shade than oaks, and as they grow in the understory, they literally shade out any new oaks that may try to grow. If left untouched, the entire forest will slowly change from oak-dominated to maple and cedar-dominated. Oak and hickory trees are fire-adapted species, while maple and cedar are not. For hundreds or even thousands of years, the forests on the Missouri Ozarks had fires that allowed the oaks to thrive and kept the maples and cedar at bay. Historically, cedar only grew on the poorest and rockiest of sites and survived due to the lack of fuel there to burn, while maple was kept to the most sheltered and moist north slopes or protected cove sites where the fires either didn't burn or burned with very low intensity.

I will say here that fire is both a friend and a foe in the forestry world. For oaks specifically, fire is a critical necessity for the continued survival of the system. On the other hand, fire can be very detrimental to wood quality if you want staves for whisky barrels or clear wood for red oak flooring. Since there are entire books being written about this and the research is expanding every day about oak forest ecology, I will just stop here and say that if you want long-term oak forests, you also need some fire.

The thing about the camp management plans we are currently working on address all these issues. We, collectively, are looking out for the long-term viability of the camp. I grew up camping and remember the oak forests; current generations are doing the same. We are setting a plan in motion so that generations 50 years and 100 years from now can also enjoy the oak forests we came to love and appreciate.

This goes hand in hand with my own situation. As Nate has been helping out the business and deciding to pursue his own forestry career, it has been fun and extremely rewarding to have his help on these projects. To watch him develop into his own rhythm as we conduct each inventory and work each follow-up project, I can see the same connection to the property that I have developing in him. It doesn't matter if we are doing plots together or dividing and conquering to get things done a bit faster; these times working together in such a special place are creating new memories for both of us.

An early photo of working with the Scouts as Mike and I presented their Tree Farm sign and Stewardship Forest Sign to Roger Hoyt.

Nate and I enjoyed a lunch at the Point with a great view.

Nate taking notes of a stand we had just finished up.

Chapter 24

No Typical Day

I get asked quite frequently what a typical day is like for a forester. Well, that answer depends. Personally, I don't feel like there is a typical day, other than the fact that we are probably dealing with trees in some fashion. Even throughout my career, there has been so much variability that calling anything typical is just funny. However, for the sake of trying to paint something of a picture concerning the happy little trees, I will give a number of examples to ponder.

Early in my career, when I was still in Blue Springs, I was sent to Eminence to help with a large timber sale marking project. This was for a full 2 weeks. Remember Coot Mountain? Yeah, that place. I have touched on that a couple of times already, so here was the typical day for that 2-week stretch. Get up and fix breakfast. Pack up a lunch of either a sandwich or leftover Casey's Pizza. Drive down the winding road from the cabin to the work site. Meet up with some of the local District staff and make sure our sounding axes and tree marking paint were all in working order. Head out and start marking trees, tallying each one. Break for lunch while sitting on a log or old stump, back to marking and tallying trees until 5 P.M..

Every other day, I would alternate between heading to the Casey's in Winona to grab a full pizza, where I would eat a couple pieces while it was hot and then take the rest back for lunches, or head straight back to the cabin to cook something, usually hot dogs, on the grill. I would then play cards with the other guy who was staying there for the same project for a couple of hours before heading to bed. The next day? Do it all over again. The weekend was a nice break, as that is when I convinced Tammy to come down for the weekend.

During fire season, a typical day may start with finishing up reports from the fires from the night before and making sure all the equipment is serviced and ready to go. There was always something to do while we waited for the first fire call to come in, and then we would spend the rest of the day and many times into the evenings fighting fires. Rinse and repeat for days on end.

It could be similar during burn season. First thing each morning, check the latest weather forecast or review the Spot Weather Forecast from the National Weather Service to see if we can burn that day. Round up all the equipment and personnel and head to the burn site. Burn all day and hope there are no wildfires to go fight and do it all again the next day.

Even today, during my consulting projects, there seem to be spurts of action. When I travel, I usually want to get the project done and get back home, so I push hard. Start the day with coffee and grab breakfast somewhere along the drive from where I am staying to the work site. Arrive and get all loaded up with water and snacks, and a lunch, and start on my plots. Many times, I eat my snacks and my lunch as I am walking between plots. I rarely sit down anymore, as my body likes to play games with me, like laughing when I want to stand back up. Work the rest of the day, always trying to get one more plot before heading back to the truck.

Depending on where I am working, supper is usually on the way back to where I am staying, unless it is with friends, and then supper is with them. Jump on the computer to type up notes from the day and then relax for an hour or so before bed. This happens day after day until that inventory is completed.

While there are these spurts of the same thing day in and day out, most other weeks could follow this pattern. Monday morning, start by reviewing or answering any phone calls or emails that came in over the weekend and line up your expected schedule for the week. Monday afternoon, go visit a landowner for an initial on-site consultation. Tuesday, go to a different landowner whom you have already visited with so you can get their forest inventory completed. As this is a smaller property, this will only take one day. Wednesday, back in the office to work on some maps and GIS work for yet another project that is coming up soon. Thursday is out and about, checking on the progress of a timber sale, checking on a thinning contractor, and visiting the local sawmill to talk about markets and prices. Friday is back in the office to start processing the inventory from Tuesday and begin to develop the forest management plan.

Even with this highly varied week, it can seem that no two weeks will be the same. The following week could start out by making a lot of site visits to homeowners in town who are seeing a sudden discoloration of leaves and leaves falling off of trees. While not a hugely critical thing, it will generate a lot of calls, so the rest of the day will be coordinating with the Public Relations people to get out a news release, schedule to talk on the local radio station about the foliar disease. Now, in addition to attending the local Tree Board Meeting that week, you also end up talking with the local Master Gardeners and the Garden Club. Oh, and just for good measure, throw in a small wildfire that needs to be suppressed one afternoon and a shipment of tree marking paint and flagging that has shown up on a pallet and needs to be unloaded and put in storage.

As a consulting forester, things can be pretty much the same in that every day can be different. Yes, I may have just spent 10 days traveling and working on two projects, but as soon as I get home, all the computer work takes place. I may spend a couple of days on GIS mapping or processing drone photos, but I can also get interrupted with incoming calls or emails from new clients. Some of these could be time-sensitive, while others can wait. I had a call once about using the drone to map a property and delineate a specific weed that was in bloom. I was getting ready to leave town for a week, so this was a very time-critical thing that we were able to do for the client.

Sometimes, I need to switch gears and look for the change. If I spend more than 2 days straight working on GIS, I tend to get a little grumpy, as that is a very tedious job and requires a lot of concentration. You kind of get into a workflow rhythm, but it can be very draining and cause a lot of eye strain from looking at a computer monitor for so long. So, sometimes, I seek out a distraction or a change of pace. A call comes in with a sick tree. Sure, I can come right over!

So, while I find it difficult to define a typical day in the life of a forester, I do also enjoy the variety. I never seem to get bored with any one task, as I seem to always have something else I can work on. Too much time outside, I can do some GIS work. Too much time on the computer, there are some sick trees to go look at. Tired of spraying paint on trees, switch to inventory, or even go fly the drone. Regardless, it is just plain rewarding that every-thing I do has to do with trees or the forest in some manner, and that is right where I need to be.

Nate and I attending the National SAF Conference in Colorado.

This is a good day to stay in and work on the computer with a warm
cup of coffee. The temperature outside was about 25 below zero,
but the trees look so pretty in their frost coating.

Getting one of our drones up for some orthomosiac development.
Photo by Tammy Shroyer

Chapter 25

I Found Nemo

I was towards the tail end of a contracted forest inventory project and was getting some help from Nate. He was on his spring break, and I was glad for not only company, but the help. This particular inventory was rather spectacularly, well, plain. Nothing really to speak of. Acres and acres of the same thing, miles and miles with no noticeable differences.

This was very odd, as usually there is something that can break up the day. Something like finding a cave, or an interesting rock formation. Maybe scaring up some deer or having an owl hoot at you. However, on this project, that all seemed to be missing. In fact, the lack of anything was starting to get weird. I mean, there were no deer tracks that we could find. There were no areas where the turkeys had been scratching. Yes, we did see a couple of squirrels, but that was it, literally, just two squirrels.

Nothing noteworthy in the trees we saw either. No unusual burls on a tree, actually, no burls at all. Even trying to take notes about the stands was getting difficult as there was nothing to separate one stand from another. Nate and I started to discuss this interesting phenomenon, noting that we both had noticed it, but took a while to bring it up. Now that it was part of the discussion, it seemed even more looming, even ominous.

On and on this went, from one plot to another. Finally, we found something different. In what looked like some animal, probably an armadillo, that was digging for a grub and scattered a little bit of dirt up on some leaves, we steered towards the "something new" sight. At last, we had something new to talk about. It looked so old, we wondered when and what animal had created this disturbance.

I opined that it was so long ago that it must have been a mammal ancestor from the Cretaceous Period, but Nate argued that, no, it looked even older. Probably the first mammal from the Triassic Period. Intermixed with our discussion about the age of this ancient animal sign we found, we also figured this was close enough to our next plot location. As was our routine, Nate had the plot center stick and started probing to find enough dirt to jam it in the ground without jarring his hands, hitting the many rocks in the area hiding under the leaves. I pulled a long strip of flagging and, using a sharpie, put the stand and plot number on it and tied it to the nearest branch while Nate did the same on a pin flag to put in the ground at the plot center stick.

I was getting my prism out of its pouch and Nate was getting his Biltmore stick pulled out of his pocket when I, just by chance, looked down a few feet away from the plot center stick, and there, half buried in the leaves and dirt, I saw something. I immediately knew what it was, but my brain could not think quick enough to comprehend what I was actually seeing. After some real mental gymnastics, I finally blurted out the words "I Found Nemo!"

Nate stopped everything he was doing and looked at me with a look of complete confusion and said, "What!?" "I found Nemo," I repeated while pointing to the ground. It was one of those "record scratch" moments for both of us. You know, on the old vinyl records, when you dropped the needle and it moved slightly and produced that loud scratching sound? Yeah, that is what we say when we have one of those moments that defy logic. The "Wait, what!?" reaction.

By now, we have both seen it. We both knew immediately what it was. I mean, the movie Finding Nemo came out in 2003, so we both had watched that movie hundreds of times, Nate by choice, me, not so much. The issue that was causing us to fight for the right words to speak was the fact that this was here. Miles away from anything. No road, and the closest trail was hundreds of yards away.

By this point, we had completely forgotten about the plot. We studied the situation, finally bent down, and fully uncovered the orange toy. How? How did this end up here? The theories started out innocent enough. It must have been dropped by a little kid. But then how did a little kid get this far from anything? Maybe they were being carried while the family was on a hike. But why were they walking on this particular slope, nowhere near a trail? Maybe a dog carried it out and dropped it here. But we hadn't seen a dog for days and hadn't heard any barking either.

At that point, the theories started to get a bit out of hand. A tornado dropped it here after destroying a town hundreds of miles away. Or, maybe it was evidence of an ancient civilization that knew about Nemo before we did, and in 2003, Pixar secretly reintroduced it to us. We finally got back on track and decided that we should get back to the plot. This proved a bit difficult as we had not fully resolved this quandary of our interesting find. Even today, the mystery has remained unsolved. Maybe we should submit it for an episode of "Unsolved Mysteries."

Alas, we did finish the plot and the inventory. But I do wonder if, instead of searching the ocean, Marlin and Dora had looked in the middle of North America, 500 miles from the nearest ocean water, and on the other side of the world from Australia, they may have found what they were looking for. I guess, maybe there is some truth to getting lost in the forest to find yourself, or at least to find something that was so lost it defies logic. Maybe it was all just a dream.

Oh well, I know the truth. I Found Nemo.

The uncovered Nemo as I first saw it on the ground.

Yep, just as I figured. I had to uncover fully to be sure.

Chapter 26

Special Moments

In looking back over the many years, I see the trees have given me a multitude of special moments and beautiful memories. I cherish each and every one of them and know beyond a shadow of a doubt that I am extremely blessed because of them. I am just going to share a bunch of them with you here. Try as I might to keep them in chronological order, I may get a few mixed up. Ready for a trip down memory lane? Here goes!

As you recall from an earlier chapter, I had a job in the basement of the Ag Building while in college. This was for a walnut study. The cool thing? Tammy also had that job; there were many times we were the only ones in the lab. As two broke college kids just trying to earn a few bucks, we were thrilled for any work, but especially work that allowed us flexibility in our hours. Those conversations in the early years of our relationship provided the groundwork for a lifetime of conversations.

Another job in college, although we were married by this point, was me being a TA for one of the two Dendrology labs. While Dr. Pallardy taught the classroom sessions, another student was to take the current Dendro classes and teach them in the field. Although I may have had some crazy antics in college, I took this job very seriously. This allowed me to take Tammy with me to prep before each lab. We would go visit the area we were to take the students to and verify the path I wanted to walk, and determine which trees we were going to cover.

These walks took us to places like Shelter Gardens, Rock Bridge State Park, Grindstone Nature Area, and Stevens Park, in addition to the many walks across the MU Campus. I have heard many times that if you really want to learn something, you need to teach it. I think that is true. The more I tried to be sure I was teaching the current students proper tree identification, the better I got as well. Tammy, tagging along on all those pre-lab walk-throughs, also learned to identify most of the trees! Those walks, with my new wife, didn't feel like work. It was just an awesome excuse to walk together, enjoy the outdoors, and teach each other new things.

In my previous book (Show Me a Firefighter), I mentioned a nice weekend spent on Coot Mountain and climbing the fire tower as the sun set. There was so much more to that weekend. I was in the middle of a large timber sale marking project, helping out, and had convinced Tammy to come down for the weekend. We took the time to visit a number of the springs in the area, like Blue Spring, Round Spring, and Alley Spring. We waded in parts of the Current River and hiked some trails. We relaxed at the towerman's house we were staying at, cooked our meals on the grill, and climbed the fire tower at sunset to enjoy that awesome view.

During some early wildfire seasons, I was gone a lot due to late nights or evening call-outs. Tammy got tired of always wondering where I was or when I would be home. So, one Saturday afternoon, we got a call for a fire out in some remote area of the Truman Corps lands. We were short-staffed to begin with, and there was a fire on a neighboring District that some of the guys were helping with, so it would be just me to fight this fire. Tammy asked if she could come along.

Well, I wasn't going to turn down the help. I quickly found an extra Nomex shirt and a spare hard hat, and away we went. We caught a pretty good break in that we could mostly back off to some trails and just burn out to secure the fire. Thus began Tammy's career in the fire service. She would go on to help out on many more fires, help with dispatching duties, assist with shuttling food, water, or equipment, whatever was needed. Her career in fire has spanned decades now and taken her across the country working on the largest and most complex wildfires while functioning at the highest levels in the Public Information Officer role.

Once, early in our marriage, Tammy and I were out for a walk through the forest. Being a forester means we don't always walk on a trail. I much prefer to just wander wherever. Well, Tammy said something about wanting to lead and go a certain direction, so I started to follow her around. As I am always looking at the trees, I wasn't really paying much attention to where we were going. As I glance at Tammy for a second and I see she is just starting to walk right over a peculiar pile of rocks. I sprint the three steps to catch her, but by now she is firmly committed, so I grab her belt and just say "Go! Go! GO!" and push, carry, will her across the rocks.

She is a bit upset with me and starts questioning what is going on. I take a survey of our legs and see we are ok. Whew! It is now that I explain she just walked right over a nest of copperhead snakes. See, she is terrified of snakes, and I was worried that if I said anything, she may freak out, hence the push to get through it. I asked what she was thinking and why she would walk over a pile of rocks like that. She just says she doesn't know; she didn't think about it. "What do you mean, you don't think about it? You have to watch out where you are going." Then she says, Well, she is used to being on a horse. Now, I just have to ask, "Well, would your horse have walked through there?" "No, he would have known better and automatically taken care of me." Yeah, some moments are special for the feeling of a myocardial infarction.

There was a time when I was working late for some reason, and Tammy and the boys brought supper to my office. As we are all sitting there, talking about who knows what, but it must have centered on the job or trees or something conservation-related and how awesome it was when Drew, being about five at the time, suddenly called out "Holy Conservation!" Of course, we all laughed, and Tammy wrote that quote down on a sticky note and dated it. Some moments are just awesome one-liners.

My mom and dad, along with one of my brothers and sister-in-law, owned a piece of property in St. Clair County, and since we were the closest, and this is what I do, we were able to help manage the property. All I asked for in return was firewood. We went down quite frequently. From doing the inventory and forest management plan, to marking out the stand boundaries, to actually cutting the trees for the thinning, we enjoyed many days there as a family. The lines between work and play were kind of fuzzy, really, as we all enjoyed the chance to be in the forest. Once, after a day of whatever, Nate grabbed the camera out of the truck and started taking pictures. Now, he was only nine at the time, but he ended up taking some cool photos. That one photo he took of Tammy and me has become a long-standing memory.

There was a time when Nate was helping me on a forest inventory for a landowner forest management plan near Hulett, WY. We worked up the ridge and came out on a large flat that provided an awesome view of Devil's Tower. We finished up a plot and took a few minutes to just take it all in. The discussion centered on the fact that we really do get to work in some beautiful places.

Actually, on that same project, but a day later, maybe, we were driving the side-by-side back and drove up on a huge flock of turkeys. When I say huge, I mean something like 60 or 70 birds, and it seemed like well over half of them were jakes or gobblers. We screeched to a stop to enjoy the sight, and I just had to play with those birds. I gobbled my best gobble, and almost in unison, every male turkey gobbled back! The sound was absolutely insane, so I did it again, and again they gobbled back. Nate and I were laughing, and he goaded me to try again. I gobbled my best gobble once more, but this time, none of them made a sound. They all just looked at me like I insulted them or something. Well, I don't really speak turkey, so I don't know for sure. Ultimately, we had a good laugh at the whole ordeal and did get a few pictures as well.

Then there was the client who offered up his cabin to stay at while we burned slash piles from a thinning project. This was for forest thinning, but also for fuel mitigation, so the trees that were cut were piled to be burned during snow cover. The fall and early winter were slow for snow in that part of the Wind River Mountains that year. We went from no or not enough snow through November and December. We took a quick consulting trip for projects back in Missouri for a week, and while there, the mountains got dumped on. By the time we got back, there was a full 24 inches of snow on the property. This was way more snow than I wanted, but we needed to get the piles burned before any more snow came.

As our client was back home in Florida, he gave us access to his Can-am with tracks to get from the highway to the cabin. Between the tracked side by side and snowshoes, we did get most of the piles lit that first day. With the short daylight, we called it a night and settled into the cabin for the evening. About midnight, we awoke to the howls of the wolves. We all got up and listened to them for a bit before heading back to bed. We were able to finish up the rest of the pile the next morning, but overall, that was a cool couple of days in the mountains.

Another time, Tammy and I were working on a forest management plan development in Missouri, and that client also offered up a hunting cabin for us to stay at while we did the forest inventory. This was a ways from town and actually a ways from the county road, so we felt like the only people for miles. Supper on the grill and a little catch-and-release fishing in his private pond made for some relaxing evenings away from the hustle and bustle.

There was the trip through North Dakota, where Tammy is from. I had a tree damage appraisal job, but we made sure to schedule a bunch of extra days on either side of that project to allow Tammy to visit family and friends. We made a stop in Mandan to visit the parents of her childhood best friend. After catching up over a cup of coffee, we walked around the yard talking about trees and the ranch. Even though they had known Tammy for her whole life, they also made me feel the same way. To see Tammy light up around them was pretty awesome. We headed on to stay a few days with her childhood best friend, Angie, and her family, at their ranch.

Hearing all the stories of her and Angie growing up, it was pretty magical to see them both riding horses across the prairie together. The time and distance of the years seemed to fade away. As this was the middle of the vast prairie of western North Dakota, there were not many trees to speak of, but they did tell me about some petrified trees on the ranch. That was good enough for me, and off I went on a long walk. After we left there, we made a stop at the North Dakota Badlands, where Tammy rode horses as a kid. Seeing her, in silence, taking it all in the sights she remembered as a kid was pretty satisfying.

As many examples as I have given, this is still just a small subset of some cool circumstances, awesome memories, great moments, and unique experiences that I have had, all due to my career as a forester. Yeah, I may have been able to grab a couple of these on a vacation or something, but just having normal days in my career that could provide all this just shows that if you love what you do, you never work a day in your life. I am happy to share some of these moments with you.

Tammy's first official fire. Due to other fires, it was just Tammy and I to fight this one.

Nate working with Devils Tower in the background.

View from the cabin with piles burning outside.

Tammy grilling supper before a little fishing.

Chapter 27

Champion Trees

I love big trees. I mean, who doesn't, right? Big trees give a sense of awe, of longevity, timelessness. What big trees are you thinking of right now? Maybe you think of the giant sequoias or the redwoods out on the West Coast that tower hundreds of feet tall. Maybe it is the wide spreading live oaks of the Southeast and Texas that can spread out well over a hundred feet. If you think of big trees in Missouri, you have probably at least heard of, and maybe even visited, The Big Tree, a bur oak in the flood plain of the Missouri River near McBaine. It even has its own Wikipedia page.

Naturally, humans like to compete, and when someone says, "Hey, I saw the biggest tree you ever saw down in the holler," someone else will invariably retort, "Yeah, but I saw an even bigger tree over yonder." I think this has happened since the beginning of time. Well, what are people to do to settle these debates? Most states have a list of the biggest trees of each species that they call the Champion Tree List. There is even a National Champion Tree Registry. Sadly, some states have stopped keeping track of their Champion Trees. But don't fret. Check out your state or even community, as some communities and cities have their own "City Champion Tree List" or something similar.

Now, if you are one to try to find a champion tree, I can give you a hint. It is actually my biggest secret. Don't look for the biggest tree. Wait, what? Yeah, I said don't look for the biggest TREE. No, find a small species, one that doesn't get real big, and then find the biggest of that species. Most of the biggest trees have already been found and nominated, so go looking for the smaller species or the less well-known species.

I used this technique throughout my career to find and nominate a number of Champion Trees for Missouri. The first Missouri State Champion tree I found and nominated was in December 1996 while I was officed at Burr Oak Woods. I was working with a landowner in Platte County and noticed a rather large ironwood tree. At least it was the biggest I had ever seen. I took some quick measurements before I left for the day. When I got back to the office, I looked up the current champion ironwood and saw that my measurements were slightly better.

I called the landowner up and told them the good news and that I would like to remeasure to be certain, and ask permission to nominate the tree. They were just as excited as I was and readily agreed. So with that, the next day I went back, dragging Larry with me, to get official measurements and some photos. Within a couple of weeks, the plaque arrived at the office. Back then, both the nominator and the owner got a plaque showing the Champion Tree name and measurements.

Not even 8 months later, I topped that current champion with an even bigger ironwood, this time over in Ray County, again on private property of a cooperator I was working with. This one was not much bigger than the first, so both were then going to be listed as co-champions. It would be a few years, and after I moved to Clinton before I would find another champion tree.

This next champion was a musclewood tree on one of the Conservation Areas I was managing. With this tree being on public land, there would only be one plaque granted to me, the nominator and finder. There is an interesting story with these first two species of champions I found, and it will take a bit of explaining.

So, what is interesting about ironwood and musclewood? Well, it comes down to the common names of trees, and why I used the names of ironwood and musclewood. First, many folks refer to these trees as American Hophornbeam and American Hornbeam. The similarities don't end there either. Let's just take each one as an individual at this point.

American Hophornbeam, with the scientific name of *Ostrya virginiana,* is also known as eastern hophornbeam or ironwood, while American Hornbeam, with the scientific name of *Carpinus caroliniana,* is also known as blue-beech, musclewood, or ironwood. They are closely related and in the same family, taxonomically speaking. They both have very hard wood, hence the ironwood name, and as well as sharing the hornbeam name, which may come from Old English, where Horn means hard and beam means wood, literally hard wood.

All this explanation of similar common names is the very reason, when talking about trees, or any plants for that matter, that you should always include the scientific name, as that name is unique and only refers to a single species. However, with all that, I just found it interesting that I was the finder of these champion trees that were so similar in so many aspects.

Remember my hint earlier? Pick a smallish tree that many aren't familiar with, correct? Yep, I looked over the state champion list and started looking for two more species, but this time with even more similarities than my first champions. I made it a mission to find a State Champion Rusty Blackhaw, *Viburnum rufidulum,* or a State Champion Blackhaw, *Viburnum prunifolium.* Of course, both of these species carry the same or similar common names and are often confused with one another. It took a few years, but I did finally find a State Champion Rusty Blackhaw down in Hickory County on the Mule Shoe Conservation Area.

There was one other champion I found while in Clinton, but I can't remember what it was. They also changed from plaques to framed certificates, and I can't find that certificate anywhere. I guess it is just lost to time, and was probably unseated by now anyway.

As it is, I don't have any current champion trees in Missouri. Those trees at the top can die or get removed, or a larger one is found, and the old one is unseated. While on a client's property in St. Clair County recently, I did find what I call a gum bumelia, or *Sideroxylon lanuginosum*. I took some rough measurements with the idea of checking the current Champion List when I got better cell service. I figured if it was close, I would go back and take official measurements and some photos and nominate it. You can imagine my disappointment when I found out Missouri had stopped, or maybe suspended, their Champion Tree Program, and not only could I not get a current list, but I couldn't even get an old list to see how close I was to yet another potential champion.

I hope that Missouri will start its Champion Tree List again in the future. I can think of no better way to get people out and about in the forests or the tree-lined parks, looking at trees. Better yet, what about getting a City or Community Champion Tree program going in your city? When I was in Clinton, we did just that. I assisted the Tree Board and the City, and we advertised the program and encouraged everyone to get out and about and start looking at those trees. We allowed any species that was on the State list as a possibility and gave instructions on how to measure. We then provided our own nomination form with the City Logo on it. We allowed an open initial nomination period, and then once we received a bunch of nominations, I and a few others went out to verify measurements. The City of Clinton, through the Tree Board, issued Certificates to the Champions. Although none of the City Champions were big enough to make the State Champion list, it was still a fun program.

As I travel through other states, I am always on the lookout for big trees, especially the smaller species. I have found a few that were large for their species, but none were ever Champion size for those states. With the ability to check your phone and pull up just about anything from the internet these days, you can see pretty quickly if a tree's measurements have potential or not.

I do have a current Champion Tree that I nominated here in Wyoming. It is a silver maple on private land within the city limits of Riverton. Doing some research on the tree and the area, my best guess is that this tree is about 115 years old. What is even more impressive is the fact that Riverton is so dry naturally, only receiving about 7 inches of precipitation annually. Trees don't grow very well here without some irrigation help. This Champion Silver Maple has had to have been the beneficiary of some well-timed irrigation for most of those 115 years.

Finally, even though they are not my champion trees, as we travel around, we are always looking for an opportunity to visit known and accessible champion trees. We have visited the National Champion Western Larch in Montana, the National Champion American Basswood in Missouri, and the newly crowned National Champion Eastern Cottonwood in Nebraska. All of these were truly big trees, and who doesn't like big trees, right?

One of my Champion Ironwood trees.

Chapter 28

Walnut

What forester who has ever worked in Missouri has not worked with walnut? Yeah, probably not many. Missouri accounts for 20 percent of all walnut lumber produced, more than anywhere else in the world, and Missouri is also the leading producer of walnut nuts and meats as well. With that kind of impact on the economy, it seems like every forester in Missouri, no matter who they work for, will be involved in learning how to manage and grade walnut trees. I was no exception.

One of the first field trainings I got from Larry on the Kansas City District was learning to grade walnuts. He had a connection with someone and was able to take me to visit a place he called the Perrin Plantation. We spent almost the entire day looking at tree after tree. Larry showed me how to be critical of the bark patterns and to "see" beneath the bark for possible defects. He taught me how to break the tree up into four quarters and judge each quarter on itself, and then to put them all together. Each grade of walnut has a metric on what size, or diameter, of the log is allowed and how many faces, or quarters, are allowed to have a defect, if any.

To give a rather simple example, a tree that might be absolutely clear on all four faces for at least 16 feet and had a diameter of 28 inches at breast height would grade out at the top level. This would be the most valuable veneer if it were harvested. On the other end of the spectrum, a tree that is only 16 inches in diameter and has a defect on three faces would not qualify as veneer at all. This would just be a lumber grade, and probably not even a very good lumber grade to boot.

Another training I got from Larry was to go with him and get a private tour of the veneer mill in Pleasant Hill. Larry, with all the right connections, arranged this for us one day. As a young forester who would be working with the trees "on the stump" out with landowners, it was very educational to see how the logs were cared for at the mill, how they were sorted based on grade, and boiled. Yes, I said boiled. Cooked. This is a process that many veneer mills use to not only soften the wood to make peeling or cutting easier, but it also helps even out the dark color that we associate with walnut, giving a more uniform color appearance to the veneer that they then slice from the log.

We then got to see the slicing machine and see the sheets of freshly cut veneer roll down the line to get stacked and carted off to the dryer. Overall, the few hours I was able to spend that morning seeing the process up close and personal helped me more than any classroom session I have ever had concerning walnut management. As a bonus that day, I was able to grab a handful of "trimmings" of some walnut burlwood that had been sliced into veneer. These were roughly square, about 6 inches to a side, and were trimmed from some larger sheets. I won't say it was scrap or waste, as there is a market for hobbyists or similar woodworkers for this sort of thing. I was thrilled to get a handful of veneer for my own projects.

Of course, there was some additional, more formal training that I received from MDC, such as from Fred Crouse, another walnut guru. Fred even helped teach me some things after he retired from MDC and was working as a consultant. Some of the state-provided training centered on assisting landowners with marketing their trees or pruning for nut production. There were agroforestry trainings where landowners could plant trees such as walnut, while still growing crops like corn or harvesting hay between the rows of trees until the trees got so big.

While walnut is still the highest potential value species of all wood harvested in Missouri, the value has changed a lot over the years. A hundred years ago, the deep, dark chocolate color of walnut wood was highly prized for furniture. Over time, consumers' taste has changed to the lighter tones of oak and now even maple or hickory; therefore, the demand has dropped and along with it, the prices. It seems that most high-grade walnuts get shipped out of the country these days, but the market for the nut meats has stayed about the same.

Most of the walnut timber sales I was involved with through my career were of smaller sales and selling just walnut for a private landowner. Some sales were as small as a dozen or so trees, but when there is veneer involved, these can still be high-dollar and therefore get a lot of interest from log buyers. Other sales may have other species mixed in, especially if all the walnuts were low quality, smaller diameter, or such a small percentage of the total sale volume.

One notable walnut sale was only eight trees. There was nothing really spectacular about them, other than that if they weren't sold, they were going to be pushed over with a dozer due to a larger irrigation pond being put in. I didn't grade any veneer in them, but did think they were some higher-grade lumber trees. I don't recall the amount they brought, but the fact that I helped them sell just eight trees is what was noteworthy.

Another sale worth mentioning is one in which we marked a total of 78 walnut trees. Over half were a bit larger in diameter, but not particularly tall. I figured and grade out some low-grade veneer on a handful, with one I graded at the highest-grade veneer. This sale received a lot of bids, with the highest bid just over $35,000!

I have personally seen a number of "prize" walnut trees in my career. Some were to stay "hidden in the back holler," according to the landowner, while others were being sold and were commanding a very nice sum of cash. One such tree, a number of foresters got to see. This was located in the fertile loess hills on the western bank of the Missouri River, up near Fort Leavenworth, Kansas. I was attending a meeting of the Missouri River Foresters, a group of foresters from Missouri, Kansas, Nebraska, Iowa, and South Dakota.

I think the meeting was actually being held there in Leavenworth, but due to travel restrictions, I was staying in a hotel on the Missouri side. Regardless, the Base Forester for the army led us on a tour where we talked about loess hills forestry, growth rates, and such. The tour culminated with the viewing of a magnificent specimen of a walnut tree. I know I have the numbers written down somewhere, but they are probably lost to Narnia as I can't come up with them at this time, so memory is going to have to do for now.

If my memory is correct, I believe this was a 40-inch diameter tree that was clear with no limbs or branches for at least 32 feet. I remember we all talking about this tree and agreeing that it was probably clear veneer on all four sides for most of that length. I also think he told us that it was sold for $20,000. It was to be cut and immediately shipped overseas. This one tree. Yeah, that was an impressive tree. I wish some of the walnuts I had sold had been that valuable.

As far as walnut for nut meats goes, people also put a lot of value into that as well. My first taste, literally, of walnuts came from my grandparents, who conveniently lived right across the street. They had a number of trees in their backyard yard and even before I knew I wanted to be a forester, I knew, and could identify about a dozen trees. I'm guessing I was 5 or 6, Grandpa Jack taught me the walnut, pecan, redbud, holly, magnolia, apple, cherry, and apricot trees that were in his yard. I also knew about the flowering dogwood and eastern white pine in their neighbor's yard. I used to climb that white pine all the time. Then there was the silver maple and the oh so fragrant black locust in our own backyard.

Back to the walnuts, I also learned early on how the husks can stain your hands. Each fall, as the walnuts started falling off the tree, I would help pick them up. I'm sure my parents were thrilled that my hands were stained with all manner of yellow and shades of brown for a few weeks each year. I watched and learned as Grandpa spread out the dried-up walnuts on a canvas tarp and then rolled them to get the hucks completely off. He then let those dry. Finally, with some gloves and a wire brush, he would clean the walnuts up before using a hammer to break them open and using small picks, pry out the nut meats. I, of course, got to help in every step of the process, but I especially like the hammer part, not the finesse of picking out the nut meats.

It is here that I will talk about the fact that walnut is NOT a good yard tree. A couple of reasons stand out, and those are the staining, as I previously mentioned, and the fact of having a yard full of balls that can either roll an ankle or get flung out of a lawnmower like a small cannon shot. Throughout my career, I have seen a number of patios, sidewalks, or streets that were perpetually stained from the walnut husks. And, while I have not rolled my ankle by stepping on a walnut, I do know people have. I do, however, have intimate, firsthand knowledge of lawnmowers and walnuts. Actually, more times than I care to admit, so I will leave that a bit vague. Let's just say one broken house window and two dented cars are all I will admit to at this time. No one saw the others or could prove anything.

Years later, I would have such a walnut tree in my own yard. I was not particularly fond of that tree due to the reasons as afore-mentioned. In addition, it was planted directly under the power-lines running through the neighborhood and had been given the traditional "Powerline Part" haircut. Maybe it was meant to be, or maybe just sheer luck, but I was home when the line-clearing contractor pulled off the highway and down our road. I stopped them, and after chatting a bit about some of my other trees that were close to the lines, I said I had a deal for them on the walnut.

"How about you prune it to the ground? Leave the main trunk right there in the yard. For any branches smaller than 8 inches, you can grind them up, but leave any larger branches, also right there in the yard. Oh, and I'll take the voucher for a new tree since you are removing this one."

See, I knew this could be a win-win. I get the tree taken care of with minimal work, and they can remove a tree that has been a lot of trouble for them. Constant trimming and pruning to keep branches and limbs from growing into the lines costs time and money, so this eliminates the problem permanently. And, I hap-pened to know about their Tree Voucher Program, so that helped. Yep, sounds good to them. With that, I left and went on to do whatever I had to do that day, and by the time I got home that afternoon, all the work was done.

I cut up the remaining branches for firewood and then winched the log onto a trailer. I know there was no veneer in this log, but it was still pretty decent for lumber, so I took it to the local Amish sawmill and sold it for a hundred bucks. "A hundred bucks? Is my tree worth that?" No. I actually get this question quite often. Nobody is going to pay you that to come and cut and haul the log away, as that just eats up their profits. I was able to do this because someone else did the work (that was beneficial work), and then I had the work and expense of loading and hauling the log to the mill myself.

I still enjoy finding those nice, high-grade walnut trees when I'm out in the forest. And, as I do a lot of work on the western side of Missouri, I still run across people who have been collecting and selling walnuts to Hammons Products in Stockton, some for multiple family generations. Honestly, if you picked up and sold walnuts anywhere from Kansas to Ohio and from Tennessee to Michigan, there is a good chance you were selling to Hammons, and those ended up in their processing plant. In the end, Missouri and walnuts go hand in hand, and I am glad I had the good fortune to be involved with some awesome experiences.

Larry grabbed my camera for me to take this photo.
Grading walnut on the stump in the Perrin Plantation.

Mike grabbed my camera to capture me talking to Becky about
the grade of this particular walnut tree.

Chapter 29

Dream Interpretation

Hey, fellow tree people. I need some help with analyzing a recurring dream.

This is the dream where we are always searching for the rare and ever-elusive Ginkgo tree with non-stinking fruit. Someone comes to tell Nate and me that they found one, so we run to find it. There are other people with us, people I know, but who I couldn't name even if I wanted to. You know, that's how dreams work, right? Anyway, there is fruit both on the ground and still in the tree, so we pick it up off the ground and start shaking branches to knock more off. But it is not the typical Ginkgo fruit; they are now buckeyes, still in the husk. So, we start frantically shucking the husk to get to the buckeyes.

Magically, a hammer appears in our hands, so we then take the buckeyes and, with that hammer, hit them to crack them open like a walnut. Inside the two halves of the walnut-buckeye-Ginkgo, we then start to pry out the "nut meat" with a small nutcracker and a pick set just like my grandpa used to have, and out pops a persimmon fruit. Now, this isn't just any persimmon fruit with all the large seeds inside; no, we split it in half by hand like an apricot and discard the single pit seed inside.

Everyone is overjoyed and eating their fill of the apricot-persimmon-walnut-buckeye-Ginkgo fruit that is the sweetest-tasting and smoothest fruit you've ever had, all the while being so excited we found one of these very rare trees.

So, who has had this dream and understands what it means and can help me out? Anyone? Anyone? No? Ok, moving on.

Illustration of my dream as done by Paige Schooner

Chapter 30

Tree Marking Gloves

Just about every forester will put paint on a tree at some point in their career. Even an Urban or Community Forester may use paint to designate a tree instead of using flagging in some situations. Other foresters who do lots of timber sales will spray hundreds of gallons of tree paint over the years. Therefore, paint splatter on your clothes, boots, tools, gloves, and hands becomes a badge of honor in a way.

Now, I have been accused of being particular, or peculiar, depending on who you talk to, but I like to categorize things. This means I like to keep my things separate when I can, and gloves are one of those things. I use leather gloves for many parts of my job, and I have a pair of gloves for each tasking. My fire gloves only get worn while fighting fires, and they smell of smoke. If I am doing something that involves the potential of getting gasoline or oil on them, I have a different pair of gloves, although these usually end up being the same at some point.

I even have a pair of leather gloves that I use when marking trees, mostly due to not wanting paint splatter on my other pairs of gloves. I really can't tell you why this bothers me so much. I mean, I like all the colors of paint that show up on my cruising vest. The more colors, the better, right? Even some of my work coats over the years contain the evidence of paint splatter from dozens of jobs, and I'm ok with that. So, I really don't fully understand my obsession with gloves and paint.

When it gets colder, I usually ditch the leather gloves for a pair of those lightweight brown jersey gloves. I just need a bit of comfort and warmth most days, and especially when doing forest inventory and needing to either write notes or punch buttons on a data collector, these jersey gloves fit the bill pretty handily. Yes, that pun was very much intended. Thank you for noticing. I usually buy my jersey gloves in the large 10-packs as I go through them quite fast. In fact, I have a number of pairs where the tip of the index finger is ripped out, so I can actually use the touch screens on our newer devices, like my tablet and iPad.

Of course, my crazy quirk also finds its way to the jersey gloves as well. Once they get paint on them, they are now permanently designated as paint gloves. Even when it warms up a bit, I am still inclined to wear gloves as I don't really like getting my hands covered in paint. Yes, it happens; it is just a part of the job. But when I can reduce the amount, then I will.

For a while, I have had pairs of orange jersey gloves for when I am using orange tree paint. These are fairly easy to find, as you just need to walk through any sporting goods store or section of a store during deer season. Personally, I think having orange gloves for deer hunting is a bit overboard, but to each their own. I'm just glad to have them and be able to find them when I need them.

The vast majority of paint that I have used over my career has been either orange or blue tree paint. Yes, I have used a little red, yellow, pink, and even green once, but even these all together only account for a drop in the bucket of my overall paint usage.

As much as I looked for blue jersey gloves, they seemed elusive, until one day I was following Tammy around as she was looking for some gloves. There was a whole women's section of gloves completely on a different aisle from the men's gloves. Lo and behold, there were some blue jersey gloves! And pink. And green. Wow, I never knew they made jersey gloves in so many colors. Well, I just had to have a pair of blue, and it really took some digging to find a pair that was big enough to fit right. These days, with everything online, a quick search of the internet will yield the colors you want, so I shouldn't ever have an excuse to be out of blue or orange jersey gloves.

Back to the early days of wearing the orange gloves, some of my crew poked fun at what I was wearing. It was all in good fun, but it was also relentless. I didn't mind, however. We were using orange paint in the gallon cans and refilling the quart cans for our tree marking guns. Someone spilled while pouring and got their nice leather gloves soaked in orange paint, while another had his hands covered in paint as well. At the end of the day, I took off my orange jersey gloves to reveal nice, clean paint-free hands. Of course, I just had to rub that in for a few minutes.

Now, once I finally got my hands on those blue gloves, the mocking started again, although this time, not as bad or for as long. Even though the blue didn't match exactly, they still served their purpose. No, I didn't get the pink or green gloves. The brown can work for those few times I use those colors.

I don't use as much tree marking paint as I used to, and I have lost my last pair of blue gloves. I still have an orange pair of gloves front and center in my truck, ready if needed, and another pair packed away in my gear. I guess I may have to search up blue jersey gloves on the internet and get some ordered again. I have a supply of blue paint that may get called on for whatever the next project is. You never know when the opportunity may present itself to splat some tree paint, and if you see me with either orange or blue jersey gloves out in the woods, feel free to say something. I can take it. In fact, I look forward to seeing what colorful insult you can come up with to poke a little fun. Just be prepared for an equally colorful retort!

Orange paint calls for orange gloves.

Chapter 31

How Far Back is Backcountry?

Growing up and working in Missouri for most of my life and career, getting an hour or 2 hours from home for a normal work project was about the extreme, and then, we may only be a mile or two at most from a paved road or the closest house. When I moved to Wyoming, my perspectives had to shift. Some projects were over three hours of drive time, and if we had any time needed for the project, it usually meant we stayed overnight in a local hotel and made a few days of it.

There was one cooperator with whom I worked who set a new record for me. This was a husband and wife who had a "cabin up the mountain" property who wanted advice on things like dying spruce, wildfire mitigation, and general forest and tree management questions. We conversed over email and a couple of phone calls, and they invited me to the property one summer day. They sent me directions, and we agreed on a day and time I would arrive. They would be there all week, and there was no cell service once I left the main paved road.

Looking at the directions and some maps, I was glad it was the middle of June when we had the longest days, almost 15 and a half hours of sunlight. This looked like it would take longer than three hours to get to their cabin. On the arranged day, I set off early, I think it was about 4:30 in the morning. An hour later, just after I had passed through Thermopolis, the sun finally peered over the horizon.

I continued on taking a left and heading due West. Eventually, the pavement ran out, and I traveled a number of miles by gravel roads. The roads seemed to get smaller and less used. I was stopping quite frequently to open gates, go through, close gates, and continue on. Down to a 2-track now, I kept referring to the directions I had printed out. Yep, there was the dilapidated old cabin, take the left fork. Then the spire rock, where I was to take the right fork.

The first of four creek crossings was a rather non-event. The track was clearly visible on both sides, and the water was very shallow, just a riffle really, then onto a seemingly random fence post, where I took yet another fork. I have to say, I think I was traveling down cattle trails at this point. As I was driving through a rather dark and shaded section of the forest, I came to the second creek crossing. I slammed on the brakes. Something didn't seem right, so I got out to take a look.

The trail turned sharply to the right to make a perpendicular crossing of the river, and it jutted up on the natural terrace running alongside the channel. Immediately where the trail entered the water, it appeared as though the water was four feet deep and freshly washed out. Hmm, I was not really sure I could go on. I could hike it from here, but it was still another three miles or so, and the trail to this point had no real options for turning around.

I grabbed a long stick and started testing the water depth. Though not four feet deep, it was deep enough to make me question if this was a good idea. After searching up the bank a few more feet, I determined I could do a multi-point turn and get a bit farther upstream before I entered the water, where it was yet a little shallower. A few white-knuckle moments and gently easing down the bank and into the water, and I was able to get crossing number two checked off. Of course, once I was on the other side, I then had the realization that I needed to get back out later today. Could I make it back up that bank with wet tires and all the water I would be bringing with me? Oh well, I guess I will find out later.

At this point, I don't remember any more forks to take. I was on the final trail. Well, at least some semblance of a trail. I hit one spot where it was a large, flat, open grassy meadow and lost the tracks completely. Rather than drive around aimlessly, I got out and stood on the top of my toolbox in the bed of the truck to get a better view, and finally saw where the trail went back into the trees, so I headed that way.

Crossing number three, and this was in my notes, involved entering the creek at a right angle and then turning to drive up the channel for about a hundred yards, and the exit from the water back to the land was at a very gradual angle. The water seemed deeper than they said it would be, and even though I had the truck locked in four-wheel drive, my tires were slipping all over the wet, rounded rocks in the riverbed.

About this time, I am really starting to wonder how anyone could get up here on a regular basis and how in the world they got supplies up this road enough to be able to build a cabin. I still have about a mile and the final, fourth crossing to go, and I was about at the limit of what anyone should be driving a full-sized truck through.

The canyon walls were pretty steep in places, and with it still being fairly early in the morning, when I was in the shadows, it was rather dark. This is prime grizzly, mountain lion, and wolf territory, so my eyes were peeled for any movement in the shadows.

Finally, after crawling at a snail's pace over a very rough trail, I reached the fourth and final crossing. This was probably the easiest one, and it meant I was home free, or maybe cabin free, as the cabin would be just around the next blind corner in the trees, and I should pop out and be at my destination.

I arrived and was greeted by the landowners. We chatted for a few minutes, and they showed me the photos of a grizzly that had walked by their front porch just an hour prior, and then showed me his tracks. It was now about 8 A.M., three and a half hours after I left, the last hour or more of some very technical, slow, bumpy driving. I was already worn out, yet my day was really just beginning.

We spent some time in the cabin looking over maps of the property and discussing bigger picture sorts of management objectives, then went outside to discuss home hardening and wildfire mitigation challenges. I took a number of photos and notes so I could develop a mitigation plan for them later. We climbed up the mountain behind their house to get a better view of the valley and where their land boundaries went. We then spent a bit of time looking at the dying trees down along the river and discussed the options they had.

Helicopter logging seemed the only viable way to get the trees gone from both a forest management and a wildfire fuels perspective. Sadly, getting a helicopter logger in that part of Wyoming, on such a small property, was not going to fly. No, really, it wasn't going to fly. Of course, the conversation did lead to how they got everything hauled up that road, which seemed more of a game trail. They just laughed, gave some wisecracks one-liners, and never really answered my question. Maybe Sasquatch had helped out.

I then asked about the river crossings, where they proceeded to tell me that each time the river gets up, the crossings change, move up or down the channel, or do something interesting. There were some rain showers the day before, so maybe crossing number two did wash out a bit. I just said, I didn't think even Jim Bridger would have tried to get this far up the canyon as they had made it. Ol Jim spent a lot of time in these parts, and he probably would be impressed at what they had accomplished.

They fed me lunch and then asked if I would like to go for a hike. Sure. I'm all about exploring. We were able to cross the creek on foot without getting wet due to some well-placed dead trees and large rocks. This put us in the middle of their largest holding of forest, and fewer dead trees, so we were able to talk about forest management options here as well. The hike would then gain a lot of elevation, and we broke out into some meadows along the ridges. Up and up we climbed, just talking and taking in the sights.

By now, we had crossed onto some BLM lands, and they then showed me some of the local archeology sites they had found. We looked and found a number of old tepee rings where the old Sheepeater People had used for summer hunting. They would erect their tepees and use large rocks to hold down the edges of their shelters to help them from blowing away in this windy environment up on the ridges. While I did take photos for myself, I won't include them here to protect the rare cultural sites.

The hike continued, and we found some elk shed antlers and even a spot where ancient cultures "mined" or dug for obsidian for making blades. Being this close to Yellowstone, there was a lot of obsidian to be found if you knew how to search for it. Alas, the day was getting away from us, and it looked like storms were brewing in the high country. Well, we were in the high country, so at least the higher country. We decided we had better start our way back with the intent that I could make it out to better roads before dark and bad weather.

After we got back to the cabin and my truck, we quickly wrapped up the day, reiterated follow-up plans I would send to them via email, and I headed back down the trail. Heading back did seem quicker than coming up, maybe due to knowing that I had been there already. It all went fairly well until crossing number two. At one point, I thought I may be spending the night out there. My first attempt to get up that bank ended with spinning wheels. I was afraid of digging down and getting really stuck, so I reversed hard and shot back across with a lot of bouncing. I was able to take a slightly different angle and got a front tire to grab on my next attempt, and was able to climb out of the river.

By the time I got back to the main gravel road at the last gate, it was well into getting dark, which is pretty late in the northern latitudes. The first sprinkles were hitting me as I got back into the truck. I was never so glad to be out of that spot as the lightning behind me was popping pretty good. I knew the river would get a little squirrelly once more rain hit it. I think I made it home that evening just after 11 pm, and thus ended an almost 19-hour workday.

I have worked in some pretty remote areas before and since that day, but I still hold that visit as the most "back country" spot of them all. I actually would like to make it back out there someday, to go on that hike again and find those bits of ancient history. Maybe I'd find a piece of obsidian to bring home, or see if they will finally tell me the secret of how they got that cabin built all those miles up that canyon. Sasquatch has been pretty tight-lipped.

Proof that a cabin does, indeed, exist at the end of the trail.

About as far back as wheeled travel is possible before the wilderness.
Lots of dead trees with no way to get them out.

Chapter 32

A Forester Wife, A Forester Life

To try to clear things up, is Tammy a Forester? The answer is a bit complicated, yes and no, to be both technically correct. Maybe by the end of this story, we can decide which answer best describes her. From the very beginning, she knew what I wanted to do and be. I knew from the age of 12, and since we started dating in high school, this has always been the plan. Looking back, I now know I was pretty lucky to have an outdoorsy kinda girl. She grew up in the prairies and badlands of North Dakota and had ridden her horses thousands of miles across wide open spaces. She knew what it was to be outside all day and how to fend for herself and survive.

Right away, she seemed to gravitate to the forestry side of things. She joined the Forestry Club with me at MU, worked the same walnut cracking job in the dungeon, and accompanied me on the many dendrology lab pre-walks. She was learning to identify trees as well as anyone, and I think she enjoyed quizzing me on the trees a little too much. She stuck with me on many side adventures where I would have to go look at a tree or visit a tree, or forest-related places. She even competed with us at the Forestry Conclave, and we did things like the Jack and Jill log roll together. We threw her into the tree identification part of the contest as well as she was the only one who had been to Minnesota, where the Conclave was being held. I mean, it made sense to all of us from Missouri.

It's funny that even years later, friends with whom we went to college still think she graduated from forestry school and are shocked when we tell them she was a PRT graduate. That's Parks, Recreation, and Tourism for those who are wondering. Anyway, she was always with me and is a large part of many of my early career stories. Since we got married the summer before our last year of college, we were still just a couple of broke, but now married, college kids. Entertainment had to be cheap, and so, it usually revolved around visiting, picnicking, or hiking in the local parks. All the while, I am spouting off the things I was learning in forestry school.

Speaking of getting married, even some of the wedding gifts were tree-centric. My grandmother gave us a quilt she had made by hand that had leaves on it. That still hangs on our wall to this day. Grandpa turned a rolling pin and made a hanger out of locally sourced wood. Yes, she still has it, and yes, she can still threaten to use it in "other" ways if I get out of line. We even got bed sheets with a leaf theme.

There was an early weekend getaway where we stayed at the Arbor Day Lodge in Nebraska City, hiked every trail they had, walked the grounds, and looked at every tree they had. As I already mentioned, there were also the weekend visits while I was on special assignments in the Ozarks, and she followed me around on fires because she got tired of not knowing where I was all the time.

Once we had kids, Tammy slowed down a bit, but not for long. When I worked with the City of Clinton and their Tree Board to create the City Champion Tree Program, she was right there, nominating trees, helping me measure other nominated trees, and calculating the winners. She helped plant the dozens of trees at our place outside of Lowry City, and although I did most of the young tree training and corrective pruning, she also jumped in when I needed help.

One summer, after I came home from a fire detail out in some western state, I was going on and on about the fun I had, the beautiful places I had seen, and the great camaraderie with other firefighters. She, on the other hand, had been dealing with two toddlers who were sick, cranky, and being cantankerous. She had to miss work when she didn't really have sick days after two maternity leaves, and she was definitely short on sleep. She, very succinctly, told me that maybe she would go on fire details, and I should stay at home with MY children. Yikes! Well, a couple of weeks later, I came home from work and told her I had enrolled her in the Public Information Officer class, and thus began her storied PIO career, where she would work on some of the most complex or noteworthy large fires in the country.

Fast forward many years, and I was now working for the state of Wyoming, and started talking about starting my own Consulting company. She was supportive as always as we decided that we should do this together. No Sole Proprietor thing would work for us. We have been partners in life so far, so we should be partners in this new company. So with that, we created FlamingTree Solutions together.

At this point, she took an even more persistent view of trees and forestry in general. While she started out helping with the books and website development, I started to take even more time to deepen the training we had already been working on for decades. We started discussing more nuanced ideas within the forestry realm, especially in regards to Urban and Community Forestry.

She joined the local Tree Board in Riverton, and her advocacy for trees grew even more. After deciding I would consult in both Wyoming and back in Missouri, it now seemed her school breaks were always spent traveling with me. Having followed me through the woods our entire life, she made a natural transition to a more skilled arborist. She has been right there beside me doing forest inventory, or tallying trees as I marked them for a timber sale. As her skills and confidence continued to grow, she started getting to do more and more things on her own.

I wanted an equal partnership, so she started to bolster her tree-related credentials by testing for and earning her Certified Arborist through the International Society of Arboriculture. Eventually, the Society of American Foresters offered the new credential of Certified Urban and Community Forester. She studied and tested and earned her CUCF credential, and I believe she may have been one of the first 100 in the country to earn this certification. While she may be humble and not speak of it much, this is a big deal and a major accomplishment, and I am extremely proud of her dedication. So, while no, she does not have a forestry degree, yes, I do consider her a Forester. Anyone who can earn the CUCF has to know a lot about trees and urban forestry.

These days, she is still right there beside me on any project. Yes, there may be some complaints about how hot or muggy it is when in the Missouri forests in June or July, or how steep the terrain is in some parts of the Ozarks or the Wind River Mountain, but she usually just tells me to shut up; I choose this life after all. Yes, I did. And I think I was on the lucky end to have the best partner a man could ever ask for as well.

At this point, I will relay a rather funny story. We were doing Community Risk Assessments in the wilds of Wyoming for a Community Wildfire Protection Plan, and stopped for lunch in one of those beautiful lunch spots that only Tammy or Nate could find. I don't remember how or why, but Tammy wandered off with her lunch in hand and ended up sitting on a rock a little ways from the truck. Nate and I, up to no good as usual, started talking about making a funny. The conversation soon turned to the old show, Wild Kingdom.

Yeah, you probably see where this is going. Without giving away what we were up to, I snuck out a bit and posed while Nate took a few photos. I was playing the part of Jim while Nate played the part of Marlin Perkins, who always seemed to narrate while Jim played the danger card. "Here we see a wild Tammy enjoying a fresh kill while sunning herself on a small rock. Be careful, Josh, wild Tammys are most unpredictable." We laughed and snickered and almost got caught.

Later, as we continued our work day, Nate and I made all sorts of references to Wild Kingdom, Jim, and Marlin. Poor Tammy didn't catch on. Well, I couldn't help myself, so we had to put that on our social media. When Tammy finally saw the pictures and figured out the whole schtick days later, she exclaimed, "That's what all the Wild Kingdom talk was about. You turds! I should have known you were up to something." It is nice to get one over on her every now and then.

Watch out Josh! Wild Tammys can be most unpredictable!
Photo by Nate Shroyer

A long way from the truck on a warm June day after we did the final check on a timber sale.

A beautiful fall day in the mountains of Wyoming after we helped a client with some thinning for wildfire fuels mitigation.

Chapter 33

Lunch Spots

I will admit that in the past few years, I have allowed myself to get into a bad habit when working alone. I really don't know what the main cause of this is. Maybe it is just working for myself, and time is money. Or, maybe, it is the fact that the quicker I get a project done, the quicker I can get home and back with the family. It could also be that I am not as young and bouncy as I used to be, and sitting on a log or rock for too long means a lot of grunting, groaning, and stretching to get up and back moving. Regardless of whatever the excuse is, I really don't like stopping for lunch in the middle of the woods.

I have found myself eating a big breakfast and then snacking all day long. Maybe grab a Nadler's snack stick from my vest and eat it while I walk to the next plot. I have gotten pretty good at eating a prepackaged Lunchables while walking as well. Sometimes, I just skip lunch altogether, figuring I will make up for it with a bigger supper. Even when I do "stop for lunch," this usually involves standing there for a minute to eat a bite, take off my vest to grab another water, eat another bite, and get back at it.

Once Nate started working with me, however, I had to adjust my eating habits. Nate still has a youthful metabolism and requires a lot more calories throughout the day than I now do. He also bounces up a bit faster from sitting on a rock or log, and with much less groaning, I must say. This also affects my prep for the day. I must now coordinate either an additional stop at the grocery store after supper, so Nate can grab a lunch for the next day, or, after breakfast, we stop so he can load up. Talk about slowing me down!

As much as I have tried to get Nate to eat and walk at the same time, he refuses to do such awful things, and instead, looks for a nice, quiet, and, many times, picturesque spot to sit for lunch. Grumbling and muttering under my breath about getting one more plot squeezed in, I have to give in and stop for lunch as well.

I will say, Nate has a knack for picking some pretty good spots. He found the right spot next to a riffle in a creek where the limestone formed a bit of a shelf, and the sound of the water was quite soothing. Being next to the creek, it was also a bit cooler and allowed us to cool off from a warm day. There was a nice shady spot on the side of a hill near a cave. While we didn't see any bats, probably due to them all being inside and napping, we were able to talk about the bats we had seen on other projects in the evenings.

There was the day we were doing the forest inventory at Bartle Scout Camp and stopped at The Point and gazed out on Truman Lake and reminisced about our times as campers and what had changed versus what had stayed the same. Of course, being that close to Iconium, it seemed as though most lunches ended up at Scott's General Store for a sandwich and a Peach Nehi Float. He took his time sipping on that float, and no matter what I said to try to get back to work, he kept saying he wasn't finished and he couldn't enjoy his float while working in the woods. Man, talk about lost productivity!

Eventually, I found a rhythm to Nate's lunch schedule. I even started trying to find the right spot to break for lunch before he did, although my success rate seems to be only about three out of ten. This is great if we were playing baseball, but not so good when Nate is keeping score.

There have been times when the weather was pretty cold, and even though I won't openly admit it, when he suggested we just eat in the truck since we were so close, I did enjoy the opportunity to warm up a bit. Let's just keep that tidbit of information between us, OK? He does not need to know that.

For the times we were working in Wyoming, there were more opportunities to grab lunch spots with some majestic views. Nate even commandeered the side-by-side one day when we were working in Northern Wyoming and took me to a spot with a better view. Goodness, who is the boss here anyway?

All in all, I guess I will say that I do rather enjoy this forced slowdown. A chance to hear the forest open up. An opportunity to have some quality father-son conversation without all the distractions. What felt like lost productivity early on has actually turned out to be some top-tier bonding time. I think I may be learning as much from him, and he is learning from me.

Nate did pick a pretty good spot. Nice and shady and cool.

Nate just said, here is where he was eating lunch and sat down.

I might take more breaks in the forest if there were seats like this everywhere. Photo by Nate Shroyer

Chapter 34

Fascinating Finds in the Forest

While I am talking about Nate, he has another interesting ability. He seems to find some really fascinating things while we are working. Even if he doesn't find them, just being around seems to allow me to find some cool things. I already talked about the time we found all those basketballs. However, we have collected such a wide array of weird things while Nate is along that I am encouraging him to write his own book solely on that subject. For that reason, I will only hit on a few here.

To start with, Nate was helping me with a forest inventory on some private land in Benton County. I would be developing a forest management plan for the client using the information from the inventory. As we were moving through the forest, Nate saw something and said we had to go investigate. Yep, an old rusted car. Now, I am not a car guy, as you can tell, so I don't have a clue what kind of car it was. Maybe you can tell by looking at the photo.

Regardless, this find derailed our workflow as we contemplated how the car got here and what the story was. After much discussion, we concluded that this must be the remains of a moonshine run. Using all of our top-notch investigative skills, we scoured the crime scene to come up with the following narrative.

At least two people were in the car and trying to evade the local law enforcement. Thinking they could lose the coppers by driving through the forest, they turned into the trees. What appears to be multiple scraps with a few trees, and the entire front of the car just disappeared, and the car rolled to a stop. That can happen when the engine is missing. Well, with the car dead in the water, or well, dead in the trees, both occupants flung their doors open and made their escape. That's our official statement, and we are sticking to it.

With that mystery solved, we then had to take a few photos so Nate could recreate the moment the driver realized his escape was foiled. With that thoroughly investigated, documented, and solved, we then got back on track for the work of the day. I don't recall any other finds from that day, but, as with the basketballs, a theme would soon start to develop.

There would be a truck frame with both axles, but nothing else, followed by some tractor parts and pieces. Following that, on a day when both Tammy and Nate were helping on another project, another car was found. We had just started our day and still had coffee in hand as we were making our way to the first stand and first plot. The car just beckoned us over to take a look. Well, you know what comes next.

Yep, we just had to investigate and, in this setting, we all pontificated our theories. It was as if we each had to outdo the other with the most outlandish story. For the sake of all involved, I had better leave those theories right there in the forest, as no good could come from sharing them. This, of course, did allow for some great laughs and another photo shoot. My favorite photo was of Nate wondering why the car would not run, and Tammy giving the "Well, there's your problem!" as she points out the obvious.

It seems like Nate is the main factor in finding weird things in the woods. Yes, I can find stuff all by myself, but it seems like I can go days, or even weeks, of working alone before I find something worth noting. I will snap a photo and text it to Nate to show him. However, it seems like when Nate is around, we can hardly go a day or two without some interesting or noteworthy discovery. In fact, if we go more than three days without finding something, we think that the stars must be out of alignment.

While there have been some themes in the finds we have, like the basketballs and now the cars, or at least wheeled vehicles, some are in a class all their own. Others only have a couple of examples. We continue to track and photograph these interesting things. Even though I contribute to the photo collection, I really do hope that Nate will someday publish his whole collection of finds and the stories that go with each.

These days, with everyone making social media content, I must admit I am a bit slow on the uptake. However, some of the finds that Nate comes across lead to some fun photo shoots. When he found a yellow turn sign along an old logging road in the middle of a project in Northeast Wyoming, we made a series of photos to tell a quick story. Since he is always up to some sort of antics, we just ran with that theme and had him turn the sign opposite the curve of the road, and the resulting photos show him "sneaking" the sign wrong, and then, "later," I come walking along and get confused. I mean, I guess I should not follow the road and instead follow the arrow and turn right? Right over the edge and down into the ravine? What is this kid trying to do to me?

We sure do have some fun with even the simplest things we come across out there in the forest. From driving the cars we find, to turning signs around, to warming our hands at the old rock fireplace, each tells its own story. I think my story is about being forced to slow down and appreciate what is out there and take the time to not only enjoy those with me, especially my family, but also this awesome natural world we live in. To not get so focused on productivity, but imagine the backstory to each interesting circumstance we encounter.

Sometimes, the most fascinating finds in the forest may be ourselves.

Tammy and Nate discussing "there's your problem!"

Nate warming his hands up in the only place we could find.

Chapter 35

The Englewood Project

In the summer of 2024, I had a wonderful opportunity to put a lot of things together. Those things were the entire forestry family, so to speak. As you have heard many times by now, my uncle Rich was a huge factor in my deciding to become a forester in the first place. Through my career, I have had the chance to work with him in a forester capacity on a handful of small projects. Then, working with Nate and all his help on a number of inventory projects, and finally, all the work I have been able to do with Tammy by my side.

As it turned out, we got a contract to complete a tree inventory on the Englewood Cemetery in Clinton. This is a two-part project where, in addition to the old school, touch every tree inventory, we were also going to do some drone imaging and multispectral analysis. Tammy and I would take the bulk of this project on ourselves, but with this being right there in Clinton, I asked my uncle if he would like to help.

Of course, he jumped at the chance. Then, with a well-timed break for Nate, we flew him in for a few days to help as well. This put our entire forestry family in one spot, on the same project. I really don't know who had more fun with this. Each of us had our own reasons for being excited. Rich had retired years ago, but with the mission to get back to looking at and measuring trees, especially with us, I saw a little extra pep in his step. Tammy, seeing the excitement on all our faces, couldn't help but grin and chuckle at us acting like kids again. Nate, getting to work with both me and Rich, soaked it all up.

I, however, will say that being able to have all of us together on the same project was just about indescribable. Being able to work on a very meaningful project with my uncle, who really turned me on to the forestry profession, was a career highlight. Seeing my son, who has come so far along as a budding forestry student and almost a forester, was an absolute "proud dad" moment. Finally, being able to share all that with my wife, life partner, business partner, and fellow urban and community forester, was just the beautiful cherry on top, so to speak.

We have written articles about being a multi-generational forestry family, and this project, where all three generations worked together, was almost emotional for me. I hope we can do it again, but even if we never get the chance to get Uncle Rich on a project again, I will hold this as one of my most cherished memories.

Over the course of five days, we identified and plotted trees, measured their diameter and height, looked for insects or disease, and made a management recommendation on almost 500 trees. Each tree was individually numbered and photographed. Yes, all in a day's work on a tree inventory, but it was who was there that made it so special to me.

All of us are together on one project. Tammy, Josh, Nate, Rich

Tammy helping with the tree inventory on the Englewood Project.

Chapter 36

Here Comes Chaos

I can't think of a better way to wrap this up. This simple little phrase has a lot of meanings for me, and I will try to hit the "big three," so to speak, and see if I can create some confusion in the very similar-sounding "big tree." Yep, see what I did there?

While chaos can be defined in a number of ways, I will start with the scientific. Forestry is a science down to its roots, so it only makes sense to build up from there. In the scientific field, chaos theory studies the unpredictable behavior of complex systems where small changes to inputs can lead to vastly different outcomes. In the world of nature, so many things are interconnected that it can seem confusing at times. We only need to look a bit further into a few of the things that have to happen in the exact sequence for a tree to grow. Now, without getting into the whole "which came first, the tree or the seed" thing, I will start with the seed.

The seed has to get to the absolute correct placement on the ground. It may fall at random and bounce into position, or maybe a bird carries it away from the parent tree but then drops it for some reason, or a squirrel grabs it up and instead of eating it right away, it gets stored for later. Regardless, through infinite possibilities, a seed ends up right where it needs to be. This is only part of the equation, as you have to look at how many seeds don't even get to this point.

With an oak tree that produces thousands of acorns, how many never fully develop due to a weevil that eats the seed from the inside? How many can get damaged by a typical summer hail storm? Those that do mature and fall from the tree are then eaten by deer, turkeys, or squirrels. The chaos involved for any one acorn to survive long enough to find the right spot seems almost incomprehensible.

Next, that seed needs the right combination of soil, temperature, and water to even germinate. Any one of these things can be off, and the seed won't germinate. Even when it does, a new combination of events is now needed to continue to grow. There has to be the correct amount of sunlight. There must be enough soil, or at least a crack in a rock, to allow for the roots to grow, and continued water in the correct amounts as well. Too much and the new tree seedling drowns, too little and it dries up.

As you can see, there are a plethora of seemingly chaotic sequences that have to happen for a seed to grow into the large, beautiful, and majestic tree we use for shade in the summer. Yet, this happens all the time, every year. We study this sequencing. We study nature. We try to not only replicate it, but also find efficiencies at any of the individual points along that sequencing. This becomes the science of trees, the science of forestry.

However, we also know that small changes along the way can lead to changes in the outcome and that we truly don't know everything we need to know about every tree, every species, and every situation. Therefore, as we manage trees and forests, there becomes a sort of nuanced dance to make the best group of decisions for the trees or forest as a whole. This is the Art of forestry.

I have always heard from older foresters, and believe myself, that true forestry is a blend of Art and Science. With as complex as the natural world is, we can strive to understand the science, but we will probably never achieve this fully. However, we also need to keep trying and moving forward, so we have to manage. These management decisions try to bring some order to the chaos, but it is not perfect, so we will keep perfecting the art until the science can either be proved or disproved.

With all that confusion so far, I will now move on to another use of chaos, and for me, this is the main trunk of the tree. If you have kids, or grandkids, or know of someone who has kids, or even if you have ever seen a kid, I think you will understand where I am going with this. Well, we had the phrase around our house, "Here comes chaos," as we would usually hear the chaos coming before we actually saw it. Nate, as most kids are prone to, usually brought with him a cloud of noise, mischief, questions, or whatever was the order, or should I say, disorder of the day. A quick example of this was stopping at a city park and seeing a playground. Of course, the begging starts to go play in said playground. It even got to logical reasoning with "what's the worst that can happen?"

Well, a mis-timed jump from Nate and a busted shin later, that produced bloodied jeans and a slow and painful hobble back to the car is what. While other examples have occurred over the years, that instance is why we still joke about "what's the worst that can happen?" when we see some adventure ahead of us.

Long story short, when Nate would come into a room, inevitably, you would hear "here comes chaos" from someone. Sometimes it even came from Nate himself. I think we were all guilty of saying it at some point, and I even believe you could see a grin form on Nate's face when he heard it. Now, depending on the day, that grin could have been sinister if he was up to something, or it would just creep onto his face with a sense of pride that he was the reason for the attention and acknowledgement.

As Nate has become so much a part of my forestry career of late, I feel it is very appropriate to include this saying. Nate's deciding to follow me in a forestry career has indeed brought some level of chaos. There were things I took for granted and thought were simple, but when Nate started asking the deeper questions, I was forced to break down my answers or rethink the "why" of it all. There were questions of why these leaves look different, even though they come from the same tree. Why did this tree grow here? Then there were the challenges of business, like why can't we add drones to our business, or why are we still doing this thing that costs us money?

To say that there has been chaos and confusion introduced is an understatement. I feel I keep up with the new science and technology of forestry, but discussing what he is learning in forestry school and comparing it to when I went a long time ago, we see that there are indeed some differences. While some of the science is still the same, like how to measure trees and species growth rates, some things have changed. One of the biggest chaotic challenges for me was learning that a handful of trees had their botanic or taxonomic family changed. As with any science, things can, and should, change when new information is discovered. In the case of maple trees, I learned them as the genus Acer in the family Aceraceae, but with recent research, they are now considered part of the Sapindaceae family. While there is too much to review as to the why, it all makes sense for the change; however, unlearning that maple trees are in the Aceraceae family and relearning that they are in the Sapindaceae family is taking some time. Talk about some mental chaos.

The more Nate works with me, the more I realize all the things I thought I knew. I thought I could outwalk anyone in the forest, until Nate grew some long legs and spent his summers fighting fires and hiking the mountains every day. I thought I understood the natural order of things in the forest until Nate came along and proposed some different, yet thought-provoking alternatives. What I once thought I had figured out and put into order has had a certain amount of chaos thrown into it, and now I am learning how to re-order this new set of challenges and knowledge.

I will end with the Biblical sense of chaos and how we should focus on our priorities. This I will liken to the tree crown with all its branches and leaves. The overarching part of it all. To begin with, I know that I was directed very early in life to work with trees, maybe even earlier than I ever realized. I have looked at this as my calling, to tend to the trees. To grow and nurture them. To cultivate and utilize them. Just as a gardener grows flowers or vegetables, or a shepherd tends to his flocks, I, too, tend to trees, and I help others do the same, whether it is a single tree in a yard or thousands of acres of forest.

God, in His very nature, brings order to the chaos, but he allows the chaos so we can learn to focus on what really matters. To each person, that may differ. What matters to me, what my priorities are, are my faith, my family, my path through life, and how I take on the responsibility I have been given.

I have long said that my favorite church is being outside, in the forest, amongst the trees. I feel a sense of being and belonging. I can sort through the mental chaos and bring some order to the confusion of life. I feel closer to my Creator when I am out tending His creation.

Family has been a priority as well. While parents and siblings are there at the beginning, even family priorities change over time as each goes their own way. Parents try their best to teach their children how to navigate this world, just as my parents did for me, and now I am trying to do the same thing. I cherish my partner in life as we do this together.

How I walk through this life, how I treat others, has also become a priority. There is too much hate, discontent, and loss in this world. I try to spread happiness and joy when I can, mostly through humor and a smile. Sometimes it is just holding the door for a stranger, or smiling because that is all I can offer at the moment. Sometimes it is kind words or words of encouragement. Regardless, I see my walk of life as a journey that we all share in some form or fashion. Some walk the same path for most of a lifetime, while others may only cross our path for a brief moment. We don't know how long anyone will be on our path, so it is wise to treat each moment as special because the moment may be gone before we want it to be.

I will include in this category of chaos reconnecting with an old friend. When you look at all the things that had to happen in the right moment for that reconnection to occur with my forester buddy Terry, some might call that a coincidence, but we call that divine intervention. As Terry will remind me yet again, there are no coincidences. Just the crafty, cultivated nature of chaos to bring us back together at just the right time for both of us.

So, yes, here comes the chaos. I am part of the chaos. I embrace the chaos. Because from the chaos comes order and priorities, and purpose. With that, thank you for allowing me to bring you on this journey to show you a bit about me, to show you this particular forester.

Show Me a Forester

This really has been one of my all time favorite photos of Tammy and I in the forest. Photo by Nate Shroyer

My favorite photo of the boys, ages 11 and 9, as they get early hands-on experience.

This timed selfie has become one of my more recent favorites of Nate and I. Smoke in the air from a recent burn we conducted, then we cut a few trees for a thinning project on a neighboring stand.

This photo speaks to thousands of words, and trees. Every acre of forest you see behind us? Yes, we inventoried every bit of that.

Reconnected by chance? Not even remotely! Remember, No coincidences.
So, what do two foresters do while cracking jokes while at the local
trout park? We feed the fish, of course! Photo with my good buddy,
Terry. Photo by Tammy Shroyer